Contents

D1762794

Acknowledgements

Even given their new capacity with CHISANBOP, the fingers of both my hands could not begin to accommodate the number of individuals who contributed to the preparation of this book. Yet to omit some, in particular, would be a gross injustice, for my gratitude to every one of them is so great.

This work really was begun some twenty years ago when CHISANBOP Hand Calculation was first conceived by Sung Jin Pai in his native Korea. A legend in his own right, by way of his mathematical achievements, it is especially notable that a mind capable of solving boggling computations without benefit of electronic reinforcement would invent a mathematics system of such ultimate simplicity that it literally is most spectacular in the hands of a child.

As this book goes to press, I still have not had the honor of meeting Sung Jin Pai personally, but I know him as surely as if we had been acquainted for years. And if one is able to judge the measure of another by his work, he is surely a giant.

It is through his son, Hang Young Pai, that CHISANBOP came to my attention. Himself a student of the method since the age of ten, the younger Pai also is a master of numbers who has continued to refine the principals of CHISANBOP Finger Calculation.

Coincidental with his arrival in this country early in 1976, my wife and I had completed the adoption of two little Korean girls who joined us in August of that year. It was through their attendance at the Korean-American School (John F. Kennedy High School, Riverdale, N.Y.) that my path crossed with that of Hang Young Pai. Our chemistry was instantaneous and shortly thereafter this journey into CHISANBOP was begun.

It is sometimes difficult to find sympathetic and understanding ears for so dynamic a venture, particularly in the realm of education, where attitudes and persuasions are as numerous as the inhabitants. For this reason, we are especially grateful to the Board of Education, Mount Vernon, New York. From my initial meeting with Mrs. Janice Mirabile Rao, Principal of Francis W. Pennington Elementary School and the searching, probing sessions with Mr. Seth Lachterman, Mathematics Coordinator—to the final demonstrations and discussions with Dr. William C. Pratella, Superintendent of Schools, together with the Principals and Math staffs of their twelve elementary schools—we experienced a singleness of attention and a stimulating enthusiasm that made this project well worth the effort and that ratified our own belief in CHISANBOP.

Prof. William E. Lamon, Ph.D., Director of Psychological Research Laboratory for Mathematics Learning at the University of Oregon, whose interest in the method has spurred the launching of a major CHISANBOP research program for handicapped, as well as normal children, worldwide.

Mr. Don Weitzman, Educational Consultant (Faculty, College of New Rochelle) was a confidant in the earliest days of the project and his words: "It's dynamic . . . move full speed ahead", were music to my then cautious ears.

I shall forever be grateful to my wife, Bernadette, who is my constant partner in this adventure, organizing the seemingly endless ingredients and providing to me a spirited, wise, witty and tireless inspiration.

E.M.L.

4

Introduction

Remember when you were in grade school and how you were reprimanded for counting on your fingers? Today, you may be continuing the tradition. But the world now has come full circle and those 10 digits at the ends of your arms are the newest, most portable calculators on the horizon. In fact, through the medium of a new system known as CHISANBOP which, transliterated from Korean, means FINGER CALCULATION METHOD, man has already demonstrated superiority to calculators. Remarkable? Indeed. And with proper practice, diligence and time anyone—adult or child, bright, average, below grade level, learning disabled or handicapped—can master the CHISANBOP Method.

While progress varies from individual to individual, depending upon such things as concentration, desire or even the need to learn the system, one absolute advantage is age . . . or should we say, rather, the lack of it? For, as with other learning challenges, such as language—including one's own—the earlier one becomes exposed, the more rapidly will the knowledge be gained. It could be said that as we get older, that which we have already learned and accepted can serve as a barrier to newer and different intelligence on the same subject. We all know how simple it is to acquire a habit yet how difficult it is to rid ourselves of it. Similarly, the younger mind which is "pure" of previously learned mathematical systems will adapt most rapidly to the CHISANBOP Method.

It is for this reason that the creator of the system emphasizes its initiation on the child of pre-school age for optimum results later. Note, however, that children beginning CHISANBOP in both primary and intermediate grades demonstrate excellent comprehension and mastery.

Teaching CHISANBOP to children at a point in their development when they just begin to grasp rudimentary number concepts will make it possible for them to escape the limited boundaries that heretofore confined them to using their fingers simply to count to 10. And before long, even complicated multi-digit problems will be solved accurately and rapidly. What is equally important is that the *meaning* of numbers will be brought home. One of the most negative aspects of the electronic calculator in the hands of a youngster is that while the "game" of pressing numbered buttons will always deliver the correct answer, the child's intellectual development and understanding of the *significance* of numbers will not have been enhanced.

We tend to believe that the calculator is our slave, yet the opposite is true. With their miraculous magnitude of computation, these "magical beasts"—in the hands of a child who has little or no comprehension of basic arithmetic—become the masters. When we fail to exercise our bodies, we become soft and flabby; when we are not required to use our minds, they too become soft. The development of CHISANBOP skills reinforces the human mind with a natural, permanent calculator, keeping it always in command.

CHISANBOP renders a larger picture by removing a child's focus of attention from the mechanical, "easy" method and by stimulating an understanding of *why* a result was achieved. While speed has merit, the opposite often is more desirable. (If one wished simply to traverse a dance floor, he could walk quickly from one end to the other.)

Yet CHISANBOP is simplicity itself. A brief practice period each day would be enough for most children to achieve a considerable level of comprehension and use. Significantly, the knowledge of CHISANBOP will enlarge a child's understanding of conventional arithmetic presently taught in U.S. schools.

How can a human system outperform the lightning output of electronic calculators? At first consideration it seems impossible. Yet, when we allow for the fact that calculators first require a human *input* of information, the delay at the *front* end of a calculation begins to become evident.

The human brain perceives images and ideas at bewildering speeds. Unlike the computer, its perception is instantaneous, requiring no intermediate function. We look at another's face and instantly our minds embrace so complete an image that we could later select that face from a crowd of thousands, even though all faces are comprised of identical components, only distinguished by their minor differences in shape and position. In the case of an electronic computer, such as a TV camera, an image must be scanned, one line at a time, until the picture becomes whole. As quickly as this happens, it is *not* instantaneous.

The speed of a calculator is only as rapid as the input of information by the human hand. Further, the calculator can only accept a single piece of data—1 digit—at a time. In other words, it cannot, like the human brain, take in 2 or more digits simultaneously.

Here is the ultimate meaning of CHISANBOP—for which the final, achievable goal is brain calculation.

Remedial Application

While CHISANBOP was not created specifically as a remedial device, it can play a remarkable role in strengthening the mathematical skills of the pupil who experiences problems with the subject.

It is for this reason that pupils well into high school have been taught CHISANBOP to improve their computational skill, their speed and accuracy and their general ability to perform. Even pupils who already enjoy good comprehension of mathematics find that CHISANBOP reinforces their skills and provides a sense of security.

Because CHISANBOP engages the pupil in a psychomotor experience, it is especially effective, partly because of its game-like quality, in better involving children who have real learning deficiencies as well as those who tend to wander or misbehave.

Not to be minimized is the aspect of CHISANBOP which renders it a constant "companion" to the user. With other manipulative devices for learning mathematics, the equipment or paraphernalia required . . . colored chips or weights or even an electronic calculator . . . must, at some point in the process, be abandoned.

One's hands, however—the only "devices" required for CHISANBOP—are always with the user, providing a continuing reinforcement throughout life. It is easy for us to comprehend the value of this most intimate aspect of the system.

Objectives of Chisanbop Finger Calculation

Those who complete this home study book will be capable of performing the following types of mathematical calculations:

1. Basic Addition: $5 + 8 = \square$

2. Multi-digit Addition (handling unlimited columns of numbers):

$$
\begin{array}{r}
23965 \\
84247 \\
56792 \\
83971 \\
+\ \ 54965 \\
\hline
\end{array}
$$

3. Basic Algebra Addition:

$$12 + \square = 28$$

Home Study Workbook

It is recommended that the reader purchase a copy of the CHISANBOP Workbook (Addition) which is carefully designed to exercise the CHISANBOP pupil in a sequence of examples that coincides with the skills presented in this book.

CHISANBOP provides one with a calculating system having a decimal structure in which the hands have a 10 to 1 ratio. In no way at odds with established methods of teaching basic arithmetic, it enhances these methods, enabling one to perform with extraordinary speed coupled with superb accuracy. The need for memorizing sums and/or products is eliminated by the ability of the pupil at all times to "find his place", both by seeing and feeling his result at any point in a calculation. Further, this unique psychomotor experience provides an element of fun which encourages the child to desire more extended participation in the "game" of learning.

Children who successfully complete the program acquire a sense of security with numbers very early in their development. CHISANBOP is notably effective in Kindergarten through 6th Grade.

Learning Sequence

Experience with CHISANBOP has taught us that while the sequence of manipulations outlined in this book is logical and natural, there are those pupils who will vary in their mastery of skills. They "catch on" to a more advanced manipulation in one category while not grasping the technique as quickly in another. It will be noted that we continually refer to Stages of progress in each of the learning areas dealt with. Ideally, there should be no rigid observance of sequence when an individual displays a comprehension of a more advanced stage.

However, a word of caution is in order. Because of the innate simplicity of CHISANBOP and the apparent speed at which progress is revealed, both the parent and the child tend to be enticed into covering too much material in any given period. It cannot be overemphasized that total comprehension of every step at every Stage of every category is essential. While it would be wrong here to fix a time of exposure for each, it should be adequate to suggest that small doses with repetition until "automatic" response is noted, would be the procedure that yields permanent knowledge and mastery of the system.

Caution
Use Only As Directed

There is a strong tendency to present the pupil with CHISANBOP skills too rapidly. You are urged to proceed at a slow pace, allowing each new skill to be learned thoroughly before proceeding to the next. Considerable time should be devoted to the skill of REPRESENTING NUMBERS. The child must be capable of quickly and accurately pressing all numbers called orally by the parent. Until this skill is mastered, one should not proceed to do calculations . . . even the most basic. It is possible that the child could require weeks to acquire this skill.

Mechanics of Chisanbop

To understand the method of CHISANBOP Finger Calculation, we must first understand the "tools" we shall be using (our fingers) and the numerical values they represent.

If we consider that each of our hands embraces a decimal system, the simplicity of CHISANBOP soon becomes evident.

Right Hand

Each finger on the Right Hand, excluding the thumb, represents a single unit (1). The right thumb represents five units (5). Therefore, when all fingers of the Right Hand are "used"—as later explained—the total will be nine units (9): 5 represented by the thumb and 4 more by the other fingers valued at 1 each.

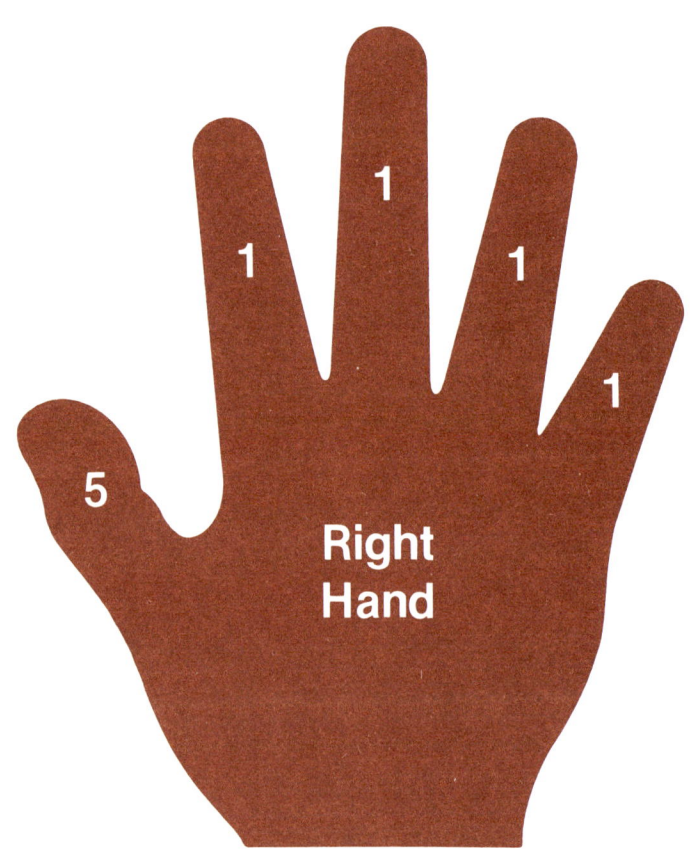

8

Left Hand

In the identical fashion, each finger on the Left Hand, excluding the thumb, represents Ten units (10). The left thumb represents Fifty units (50). Therefore, when all fingers of the Left Hand are "used"—as later explained—the total will be Ninety units (90): 50 represented by the thumb and 40 more by the other fingers, valued at 10 each.

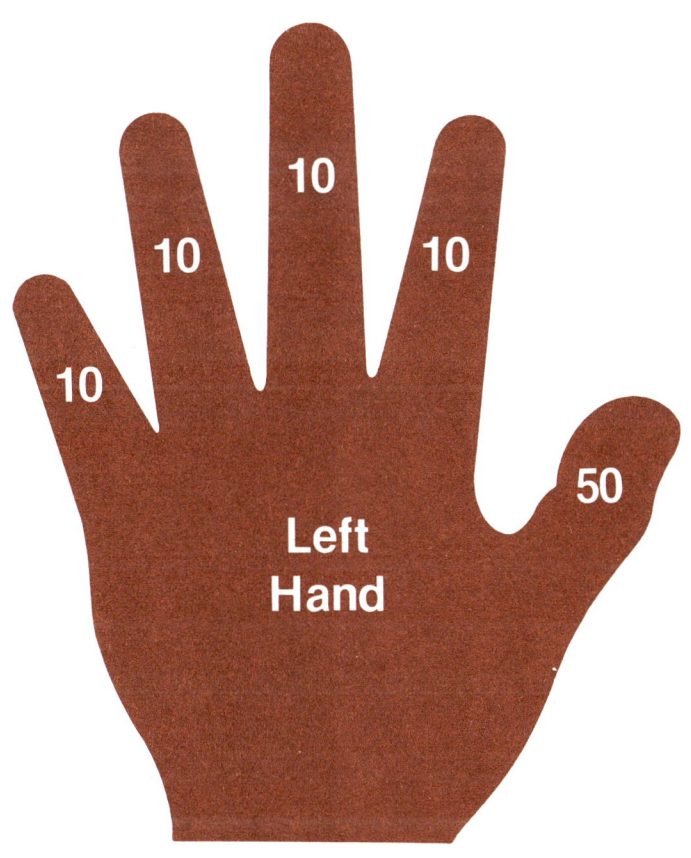

Both Hands

Therefore, combining the values of both hands, it is evident that we can now use our ten fingers to achieve a total count of 99 (9 on the Right Hand and 90 on the Left Hand.) We have 99 fingers which are capable of accumulating, storing and calculating!

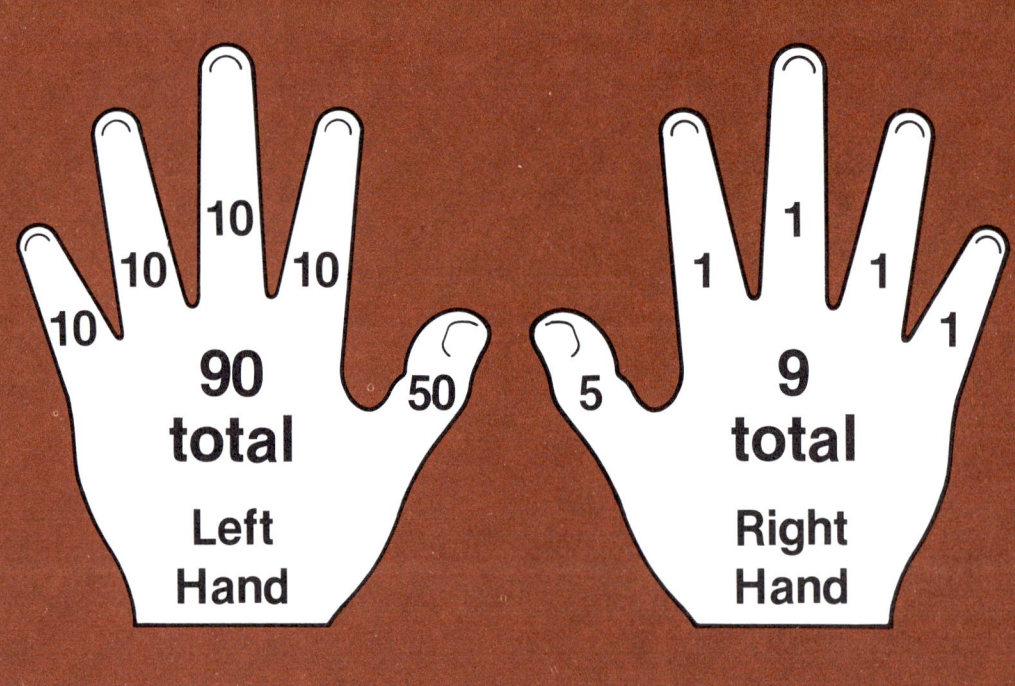

How to Use our Hands

In the initial stages of CHISANBOP there is considerable reliance on psychomotor activity. That is, the action of one's fingers and the positions they assume become a real support to what would otherwise be a purely mental activity. Further, it will become clear as we progress with this explanation that our fingers can be read, as surely as if numbers were appearing on a screen. Therefore, one is capable of *seeing* and *feeling* his answer and knowing his "place" at any given point in a calculation. Ultimately, the physical manipulation of fingers is abandoned as the technique becomes internalized.

To gain this preeminent manipulative experience, the position of one's hands becomes essential. *(Refer to illustration on Page 78)*

Imagine you wish to play the piano with your Right Hand. Instead of a keyboard, however, a desk or table will be used. Just as you would suspend your hand above the piano keys and then strike one or more keys, as desired, so will you strike (or Press) the desk surface with one or more fingers, as required by the numbers you wish to establish. Fingers are to be kept in a relaxed, spread position for proper manipulations. To achieve properly arched hand position, arms should be kept parallel to desk surface. Fingers should never be curled under but always "open" for instant use.

Symbols, Diagrams and Instructional Terms

Hereafter, we shall use the term "PRESS" and the symbol (▽) to mean that a particular finger (or combination of fingers) will make contact with the desk surface. The opposite action will use the term "CLEAR" and the symbol (▲).

In both instances the diagrams used in this book further reinforce an understanding of the kind of finger manipulation being employed. A finger being activated into a PRESS position will be shaded and have the symbol (▽) above it.

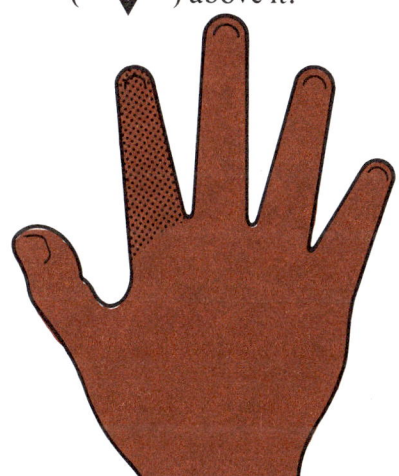

A finger being activated from a PRESS position to a CLEAR position will be clear and have the symbol (▲) above it.

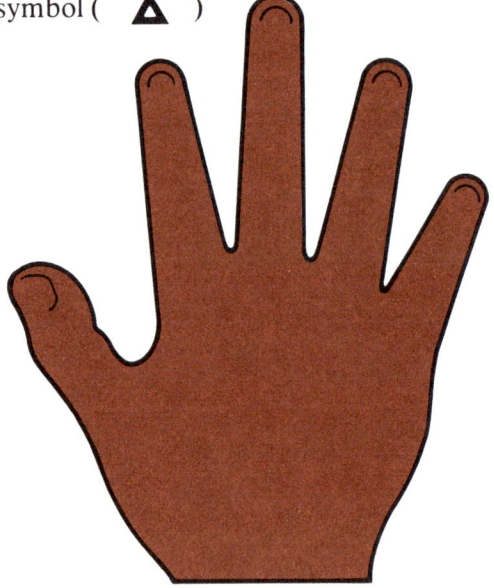

A finger which was previously PRESSed and is to remain PRESSed (while other fingers may be activated) will be black in color and have *no symbol*.

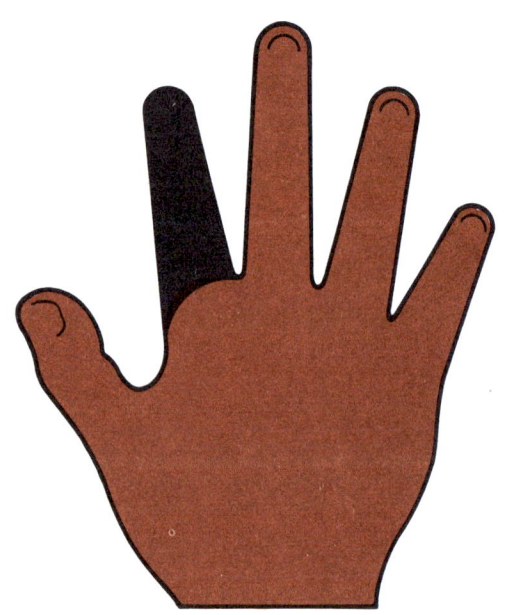

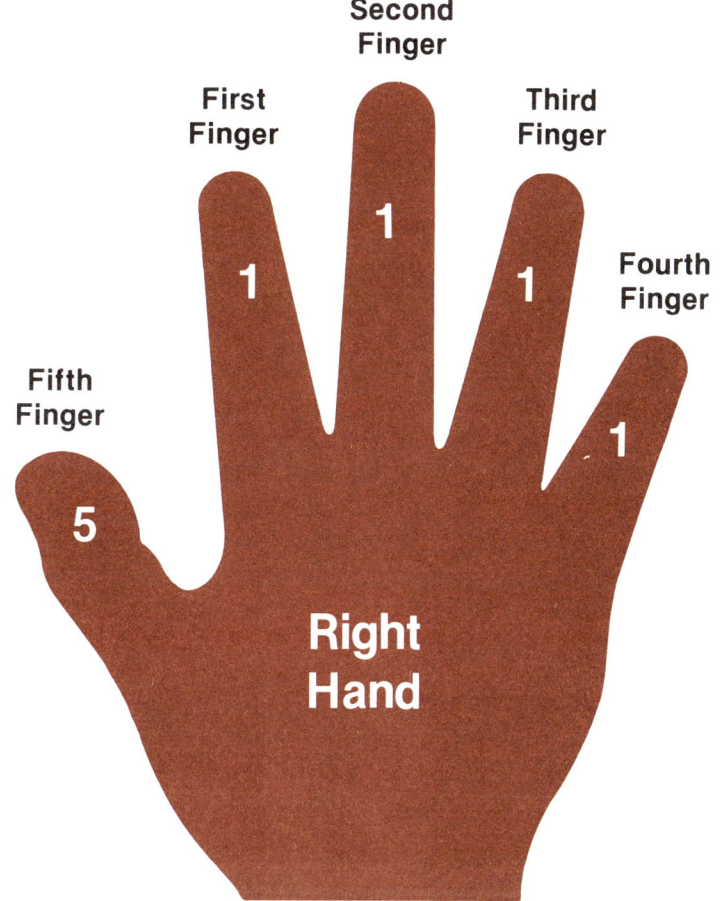

Naming the Fingers—Right Hand

As stated earlier, the four fingers of the *Right Hand* (not including the thumb) each has the same value, namely 1. For simplicity, rather than refer to each finger by its common name (Index Finger, Middle Finger, Ring Finger and Pinky), we shall call them (from left to right) First, Second, Third and Fourth. The thumb will be called the Fifth Finger.

(It is important not to confuse the "name" of the finger with its numerical value. That is, the Second, Third and Fourth Fingers *still* only have a value of 1 each).

Representing Numbers

We are ready at this point to establish all of the numbers of the Right Hand.

Parent: **We know how to say the numbers from 1 to 10. They are** (write them):
1, 2, 3, 4, 5, 6, 7, 8, 9, 10. Let's say them together. Ready? Begin! (The child counts as you point to each number)

Child: 1, 2, 3, 4, 5, 6, 7, 8, 9, 10.

Parent: **Very good. Now I'd like to hear you do the same thing by yourself.** (Child repeats above exercise.)

Very good. Now, I'm going to teach you to do something very special. You are going to learn how to use your fingers to count numbers and to do arithmetic. First, I want you to Press the fingers of your Right Hand to your desk like this: (Sit next to child and Press all Right Hand fingers to the desk as shown below.)

Have you got ALL your fingers Pressed to the desk? (sit beside the desk, help to adjust the child's hand.) **That's very good. Now, I want you to remember that when I say 'PRESS', it means to Press your fingers to the desk . . . just the way you're doing now. And when I say CLEAR YOUR FINGERS it means to raise them over your desk, like this.** (Demonstrate, alternately explaining:) **PRESS . . .CLEAR YOUR FINGERS . . . PRESS . . . CLEAR YOUR FINGERS. Good. Now let's do it together once more, like this: PRESS and CLEAR YOUR FINGERS.** (Repeat 3 times.) **That's very good. Now, we're ready to learn some numbers. We'll begin with Number 1.** (Write **1** on paper) **Watch what I'm doing.** (At desk, Right Hand is raised, ready to Press First Finger.) **This is the First Finger.** (Point First Finger to ceiling) **When I say PRESS ONE, you do this:** (Press First Finger to desk, keeping other fingers suspended.)

NUMBER 1 **PRESS ONE!** (First Finger) (keep other fingers suspended)

Ready now, let's do it. PRESS ONE. Keep your other fingers off the desk. Only PRESS your FIRST finger. Very good. Now . . . CLEAR YOUR FINGERS. (Child raises fingers from desk.) **Let's do it again. Ready? PRESS ONE.** (Child does.) **Now, CLEAR YOUR FINGERS.** (Child does.) (Repeat 3 times.) **Very good. Now we'll try the Number Two.** (Write 2 on paper.)

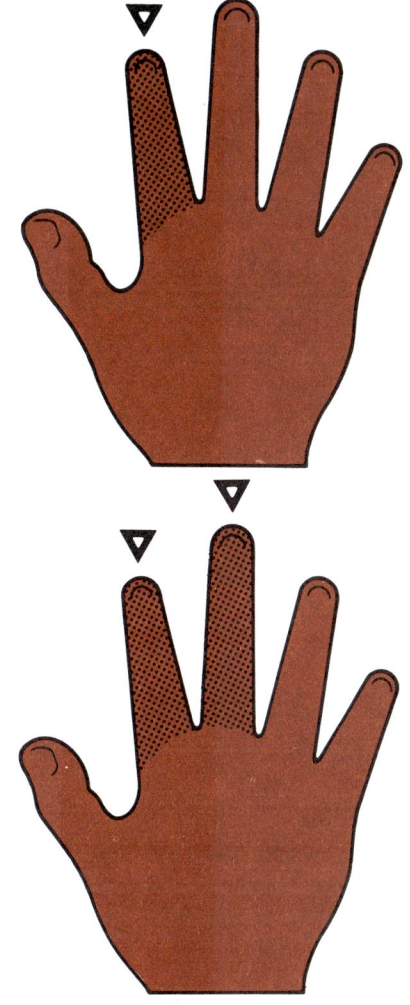

NUMBER 2 **PRESS TWO!** (First and Second Fingers) to the desk in sequence.

First, we PRESS ONE. Then . . . watch me . . . we PRESS ONE MORE. See how I use my Second finger? (Demonstrate.) **ONE and TWO. ONE and TWO. Now, let's do it together. First, CLEAR YOUR FINGERS. Ready? PRESS TWO! ONE and TWO. That's good. Let's do it again. CLEAR YOUR FINGERS.** (Repeat exercise, with child calling out "One, Two" . . . 3 times.)

NOTE: Beginners of CHISANBOP Finger Calculation in the *First Stage* will require this finger-by-finger function when going beyond the Number One. That is, on the instruction "Press Two", instead of Pressing the First and Second Fingers of the Right Hand simultaneously they must be Pressed one at a time in sequence. The pupil Presses the First Finger exclaiming "One" and then while still pressing this First Finger, will press the Second Finger, exclaiming "Two". In the *Second Stage* this intermediate function will be abandoned. Then, instead, on the instruction "Press Two" the First and Second Fingers will be Pressed simultaneously, with the single exclamation "Two!" In the same manner, the Number Three and then the Number Four will be achieved in the First Stage by pressing, in sequence, the First,

Second and Third Fingers and then the Fourth Finger, exclaiming "One—Two—Three" and "Four". To reach the Number Five, repeat the sequence to Four and then Press the Right Thumb while simultaneously Clearing the other four fingers, exclaiming "Five". (See p. 17 "Thumb Has Dual Function".) The human hand provides a natural separation between the thumb and other fingers which makes this transition from Four to Five simple to achieve.

It must be emphasized here that the reader must not be tempted to forego the exercise of Pressing individual fingers in sequence while counting orally. It is through such practice that proper preparation is made for the Second Stage technique of Pressing fingers simultaneously.

For each of the Numbers 3, 4, 5, 6, 7, 8 and 9, use the same pattern of dialogue as outlined earlier for Number 1 and Number 2, when working with a child.

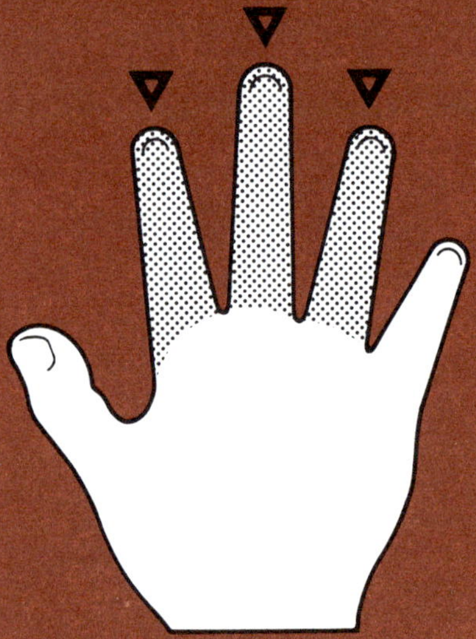

NUMBER **3**

PRESS THREE! (First, Second and Third Fingers) in sequence.

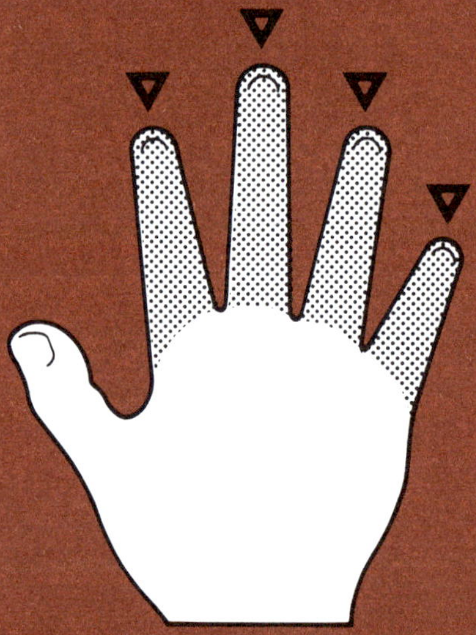

NUMBER **4**

PRESS FOUR! (First, Second, Third and Fourth Fingers) in sequence.

Thumb has Dual Function

The very essence of CHISANBOP is in its provision for a Right Hand capacity of **9** and a Left Hand capacity of **90**. This is made possible by the dual function of each thumb, which has both an ordinal and a cardinal value.

We count in sequence on the Right Hand, starting with Number 1, represented by the First Finger and continuing with the Second, Third and Fourth Fingers to the count of 4. Then we Press The Fifth Finger (Thumb) for the count of 5. At this instant, we EXCHANGE the Thumb for the other four fingers which are taken away (Cleared). This accomplishes two things:

A. The count of 5 is established (via the addition of a single finger: the Thumb). This Thumb represents the articulation of "5" as well as that of a new unit.

B. The First, Second, Third and Fourth Fingers are Cleared to be used again in the progression from 6 to 9, simply by adding them one at a time to the now established "5". Were these four fingers to remain Pressed when the Thumb was initially Pressed for the count of 5, our numerical limit would have been reached.

This dual function of the Thumb as Count Articulator (the First Finger in a +1 sequence after the flow reaches the Fourth Finger) *and* as a 5 Unit (exchanged), gives CHISANBOP a calculator-like mechanism for sequential and set accumulation.

The identical Exchange reveals itself on the Left Hand, as later illustrated, where the Thumb represents 50 and the other fingers each represent 10. What could otherwise be a confusing element of the system is simplified by the natural physical separation of the Thumb from the other four fingers.

NUMBER **5** PRESS FIVE!

(Proceed from Number 4 by Pressing Fifth Finger and simultaneously Clearing the First, Second, Third and Fourth Fingers).

NOTE: On reaching the Second Stage the pupil will abandon the finger-by-finger progression from 1 to 5 and simply Press the Fifth Finger (5) on the instruction ''PRESS FIVE!''.

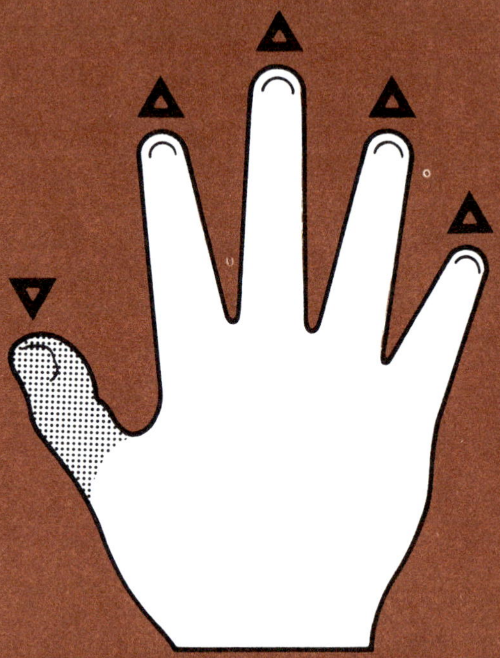

NUMBER **6** PRESS SIX! (Fifth and First Fingers) in sequence.

Once again, in the First Stage one must first use the finger-by-finger method to reach the Number 6, counting orally as each finger is Pressed: ''One, Two, Three, Four, Five, Six''. In the Second Stage, pupil Presses Fifth Finger (5) saying ''Five'' and then Presses the First Finger saying ''Six''. In the Third Stage, one simultaneously Presses Fifth and First Fingers, saying ''Six''.

This same progression of Stages will also be used for the Numbers 7, 8 and 9.

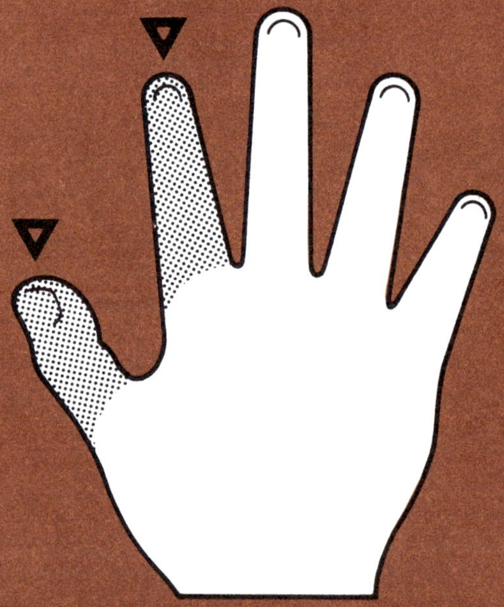

 18

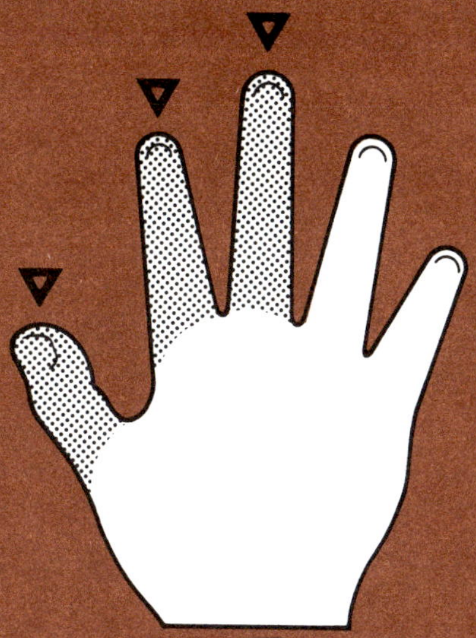

NUMBER 7 **PRESS SEVEN!** (Fifth, First and Second Fingers) in sequence.

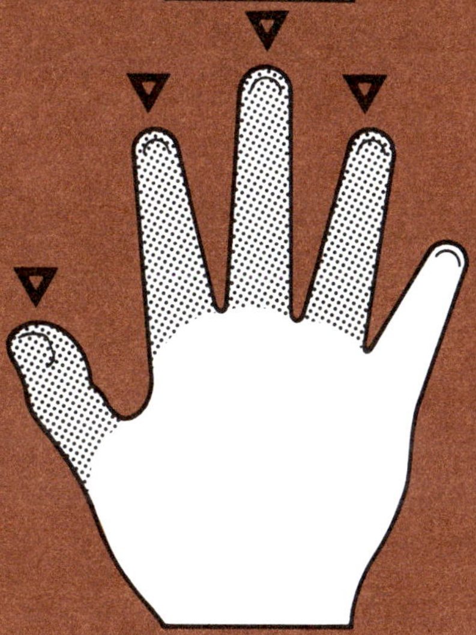

NUMBER 8 **PRESS EIGHT!** (Fifth, First, Second and Third Fingers) in sequence.

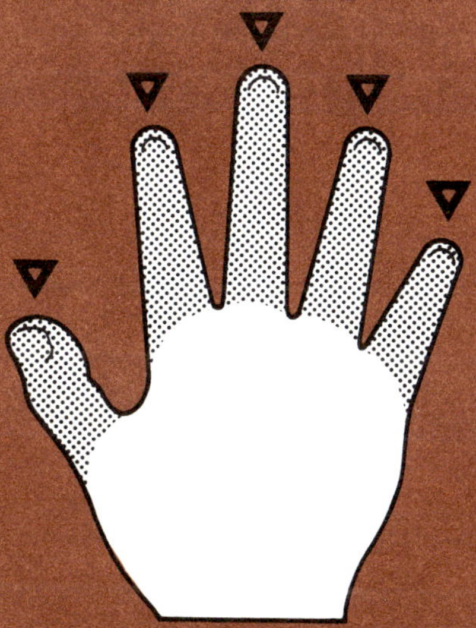

NUMBER 9 **PRESS NINE!** (Fifth, Second, Third and Fourth Fingers) in sequence.

This completes the numerical capability of the Right Hand, with which we can Press any number from 1 through 9.

 19

Establishing Numbers from 1 to 9

One must practice this fingering in sequence, Pressing the Numbers from 1 to 9 and saying the numbers out loud simultaneously, thus establishing a correlation between the physical experience and the number being expressed.

It is also essential to exercise the progression of fingers from 1 to 9, while saying aloud:

"One, and one more . . . and one more . . . and one more . . . and one more . . . and one more . . . and one more . . . and one more . . . and one more."

This establishes the awareness that 5 Ones make 5 and that 9 Ones make 9, which is of increasing value later when the subject of adding numbers is approached. It is a perfect example of why a knowledge of the *meaning* of numbers is lost when using an electronic calculator where the single unit approach is bypassed. To add 4 to 1 on a calculator, one simply enters 1 + 4, arriving at "5" but having no awareness of how this sum was achieved.

As stated earlier, depending on the child's command of the more elementary function of finger-by-finger progression, the parent will then progress to a higher Stage calling out the following random sequence, to which the child responds by Pressing the necessary fingers *simultaneously* (no longer using the single finger progression) and expressing the number orally.

PRESS ONE . . . CLEAR YOUR FINGERS . . .
PRESS FOUR . . . CLEAR YOUR FINGERS . . .
PRESS EIGHT . . . CLEAR YOUR FINGERS . . .
PRESS FIVE . . . CLEAR YOUR FINGERS . . .
PRESS NINE . . . CLEAR YOUR FINGERS . . .
PRESS THREE . . . CLEAR YOUR FINGERS . . .
PRESS SEVEN . . . CLEAR YOUR FINGERS . . .
PRESS TWO . . . CLEAR YOUR FINGERS . . .
PRESS SIX . . . CLEAR YOUR FINGERS . . .

The child will respond by simultaneously Pressing the required fingers and expressing the number orally.

It is essential for the child to learn the value of each finger, recognizing it without hesitation both by sight and by touch. Of equal importance is the ability to read these values in combination with one another. By concentrating first on a keen awareness of the Numbers 1, 2, 3, 4, 5, 6, 7, 8, and 9 . . . being able to Press each number instantly on random instruction . . . and then by concentrating on a keen awareness of the Numbers 10, 20, 30, 40, 50, 60, 70, 80, and 90 (as explained in the next section describing Left Hand values), also being able to Press each number instantly—even when called in a random fashion, the youngest child will learn quickly to Press combinations of the two groups to establish intermediate numbers.

Because the Numbers 1 through 9 are exactly the same, no matter which Tens number they are used with, it is simple for one to learn these combined values.

The only exception is when one must establish any number between 11 and 19, where the use of the words "One, Two, Three, Four, Five, Six, Seven, Eight, Nine" are not appended to the Tens number as they would be with 20, 30, 40, 50, 60, 70, 80, and 90. This is a real barrier to the very young child in his effort to learn the decimal structure. (Ideally, the numbers between 10 and 20 could be renamed "Ten-one, Ten-two, Ten-three, Ten-four, Ten-five, Ten-six, Ten-seven, Ten-eight, Ten-nine . . . to make initial comprehension of this sequence easier.)

No attempt should be made to progress to the subject of "How to Calculate" until a complete mastery of Finger Values is established.

This is the single most important skill to be mastered, if one is to become competent with calculations.

Naming the Fingers—Left Hand

When comprehension of the Numbers 1 through 9 is noted, proceed to the Number 10, which requires a function of the Left Hand.

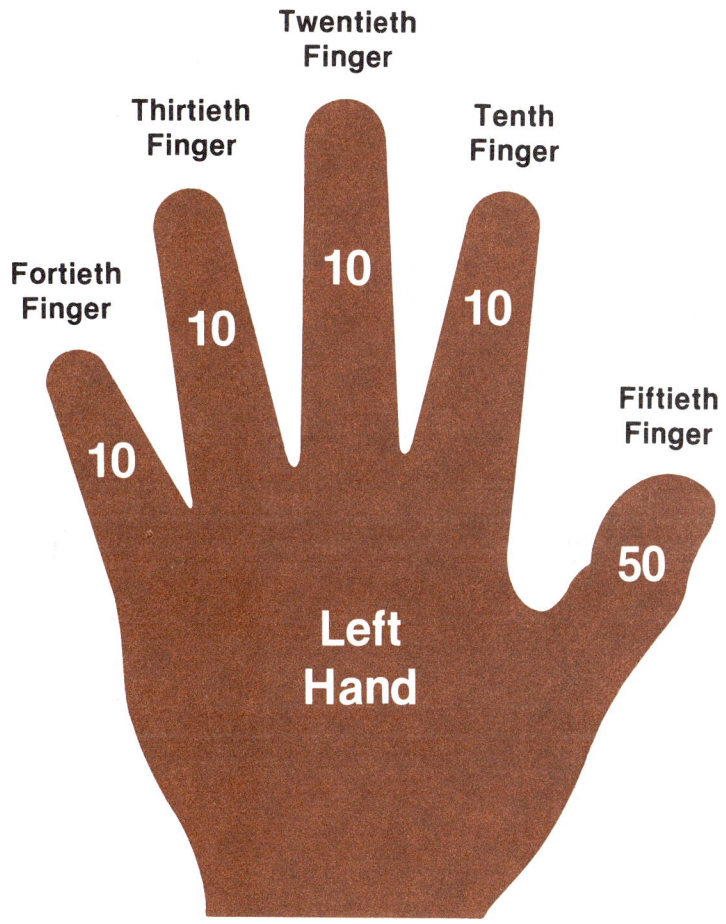

NUMBER 10

PRESS TEN! (Tenth Finger) *only*. (Do not proceed any further at this point. Return to the Right Hand sequence of Numbers 1 through 9 and then go directly on to 10 (Tenth Finger) on the Left Hand. Note that this is a major juncture, requiring the simultaneous Pressing of 10 on the Left Hand and the Clearing of the entire Right Hand. This is identical in principal to the earlier EXCHANGE on the Right Hand from 4 to 5, where the first 4 fingers were EXCHANGED for the Fifth (Thumb). Here, to achieve 10, the Tenth Finger is Exchanged for the entire Right Hand which has accumulated a total count of 9. To allow the Right Hand Fingers to remain Pressed would represent 9 plus the 10 of the Left Hand—or 19! This Exchange now Clears the fingers of the Right Hand for further use—with a *new* capacity of 9.

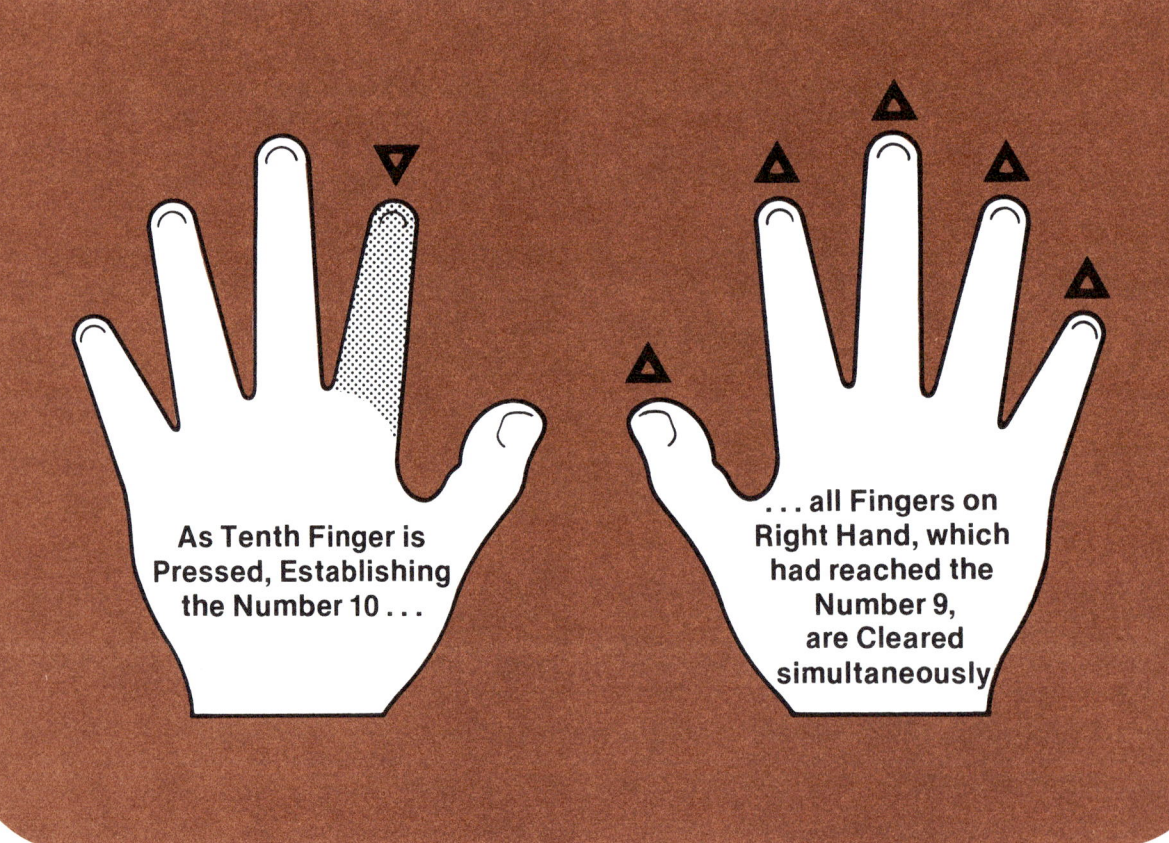

As Tenth Finger is Pressed, Establishing the Number 10 . . .

. . . all Fingers on Right Hand, which had reached the Number 9, are Cleared simultaneously

Decimal Pattern
of
Chisanbop
Finger Calculation

The Exchange
from
Units to Tens

Exchange from Units to Tens
(A Decimal Pattern)

When counting consecutive numbers in sequence, it will later be noted on reaching 20, 30, 40, 50, 60, 70, 80 and 90, that a Left Hand Finger is always Pressed at the same time the entire Right Hand is Cleared. As it is from 9 to 10, so will it be from 19 to 20, 29 to 30, 39 to 40, et al. Once again, an Exchange takes place wherein all the fingers of the Right Hand (total of 9) are Exchanged for the upcoming TENS Finger of the Left Hand.

A child's ability to count orally to 10 or 20 or 100 will be helpful in comprehending the above. Other than with the Numbers 11 through 19, our language repeats—from 20 on, the words "One, Two, Three, Four, Five, Six, Seven, Eight, Nine", so the transition from Right Hand to Left Hand always embraces this decimal pattern of 1 to 9 on the Right Hand and then units of 10 on the Left Hand.

Being aware of the child's existing knowledge of numbers, by hearing him count orally in sequence, the parent can gauge the level at which he is capable of learning the CHISANBOP Method.

NUMBER 11

(Here, for the first time, we call upon the use of both hands to establish one number.)
PRESS TEN! (Tenth Finger) **AND PRESS ONE MORE** (First Finger) in sequence.

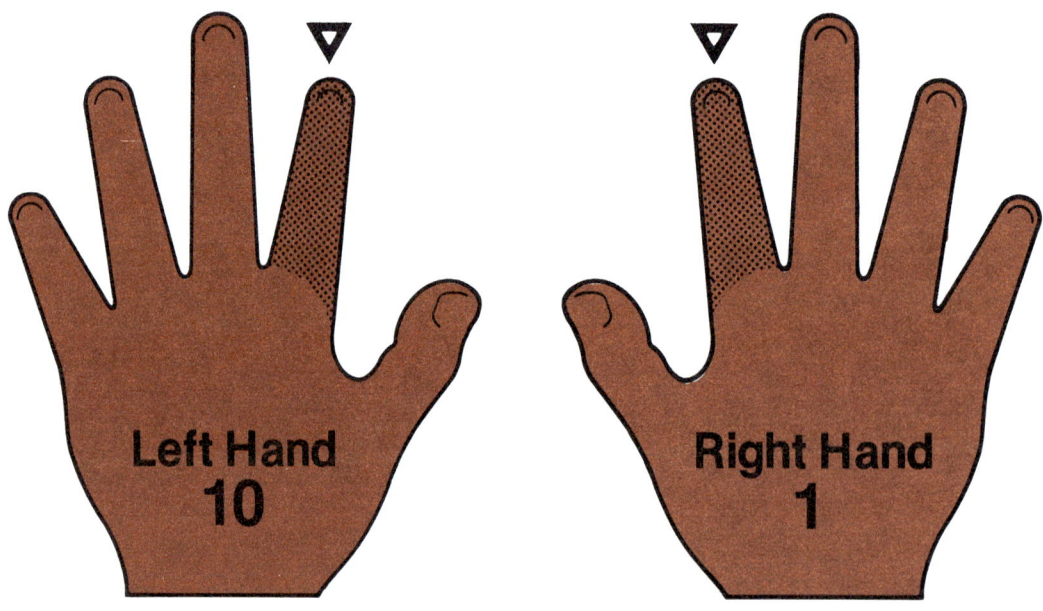

Left Hand
10

Right Hand
1

Once again, it is essential to practice Pressing each Finger, *one at a time*, beginning with One—and counting orally—until the count of 10, and then 11 is reached. Subsequently, this finger-by-finger function will be replaced by the single step procedure which is to Press 11 (Tenth Finger and First Finger simultaneously).

Note at this point that this reading of fingers is not to be confused with the later technique of adding different values to one another. That is, in the first instance, the child comes to recognize that this combination

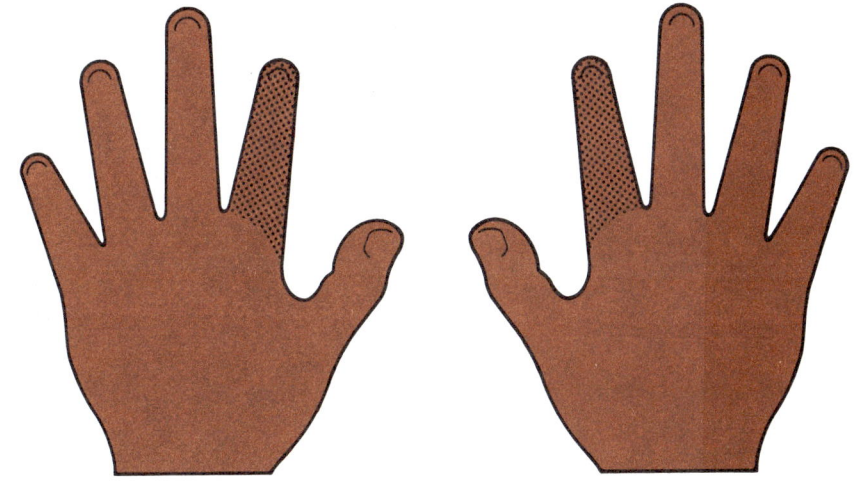

has a value of 11. It *is* Number 11. Later, when the technique for Addition is learned, the child will understand that

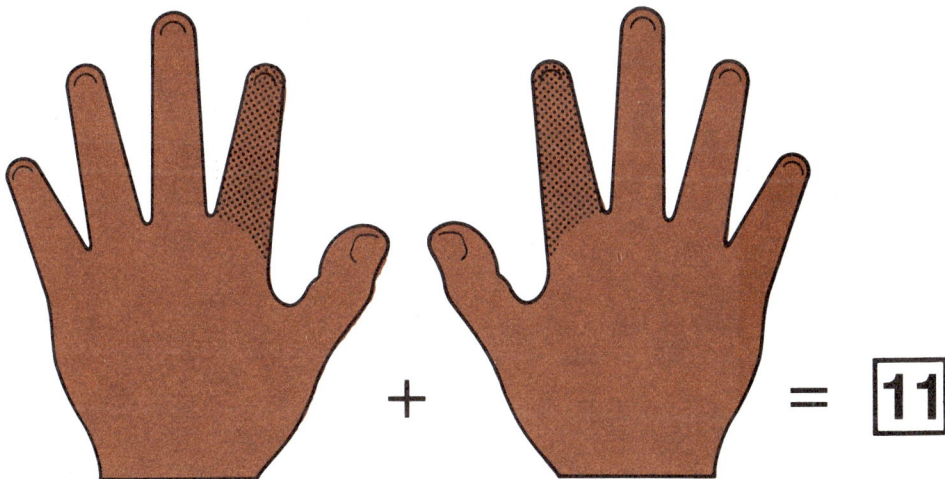

There is a subtle difference but a significant one, in that the child *will* comprehend the reading of a Press (One hand or Two hands) much before the logic of Addition is understood.

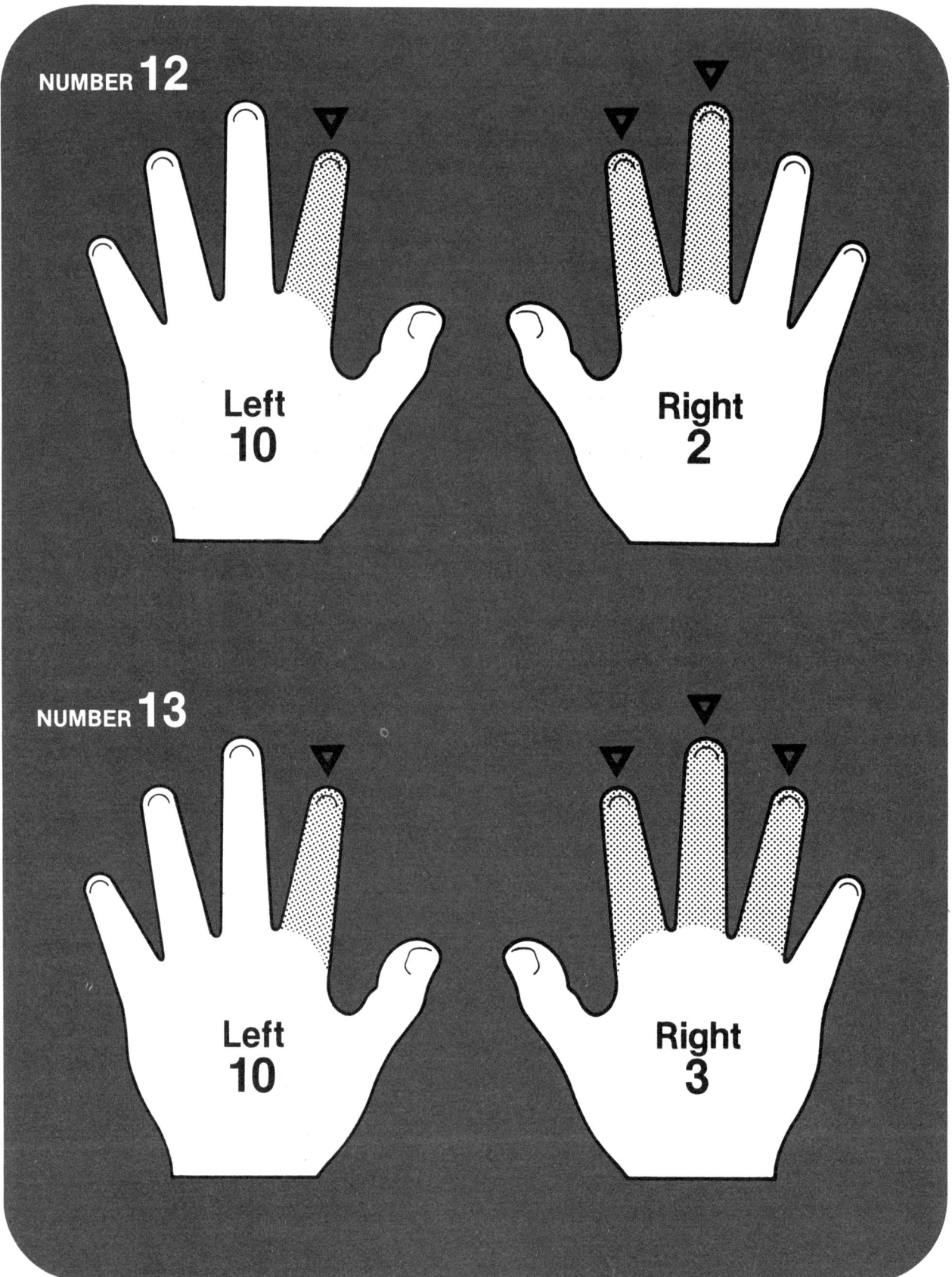

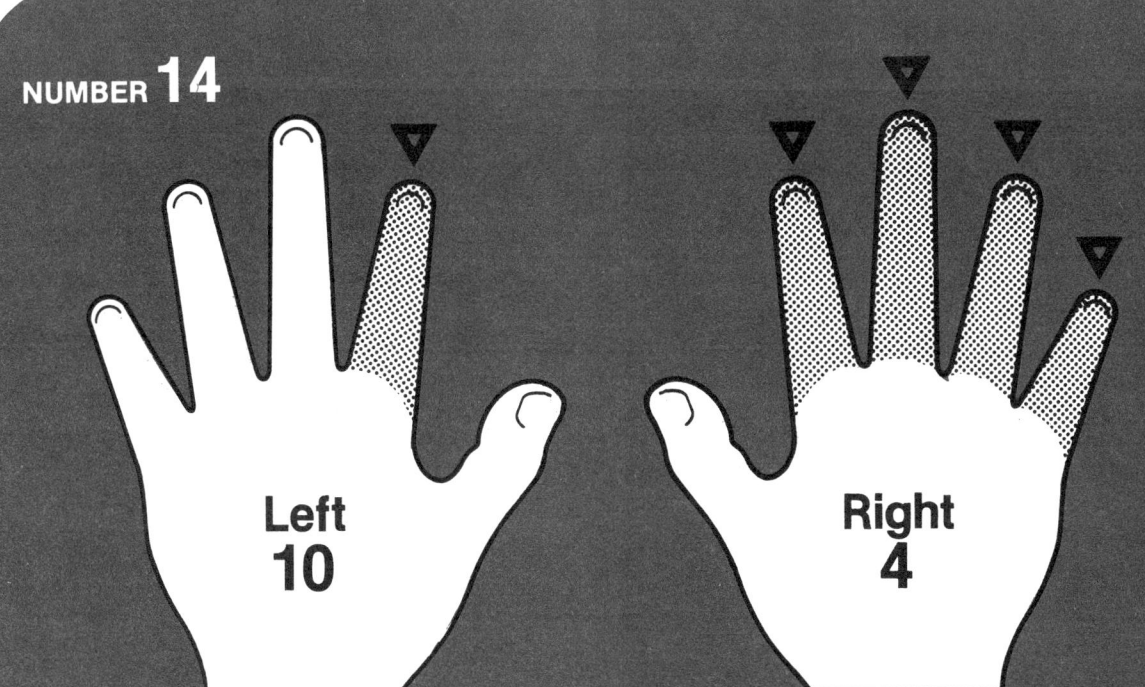

NUMBER 14

Left 10

Right 4

After establishing 14, as illustrated, be prepared for Right Hand Exchange of Pressed Fingers for Thumb (5) . . . to achieve 15.

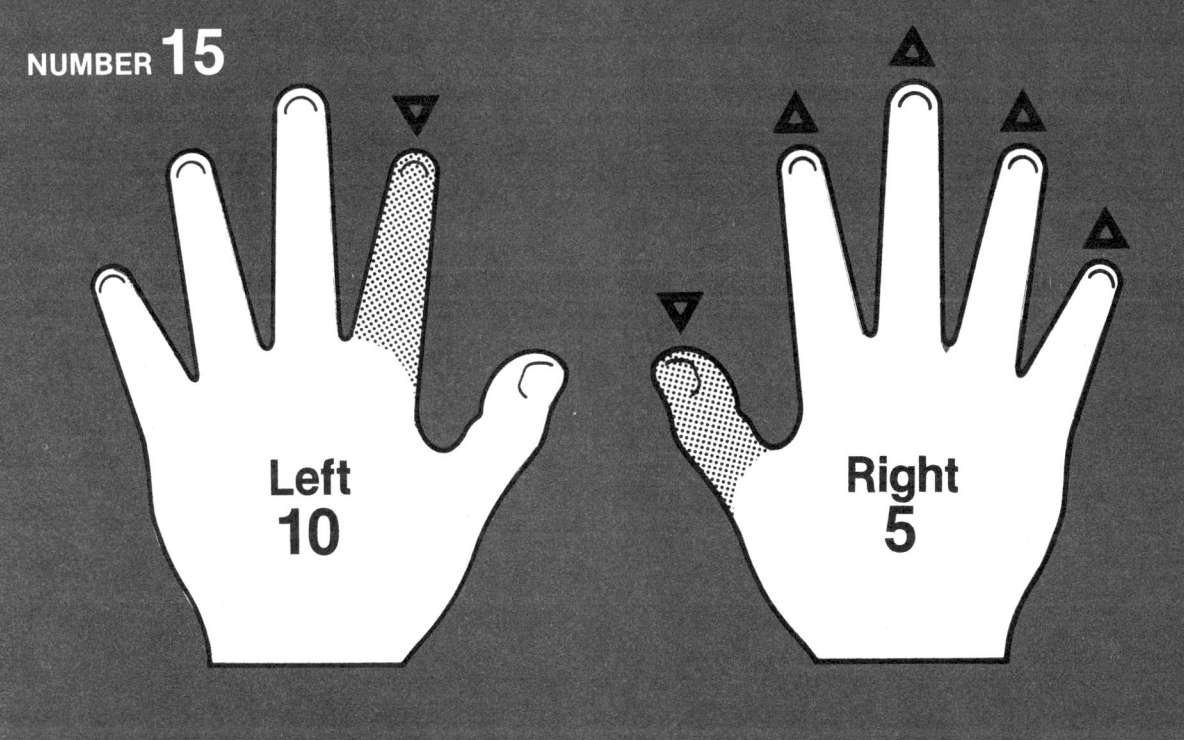

NUMBER 15

Left 10

Right 5

NUMBER **16**

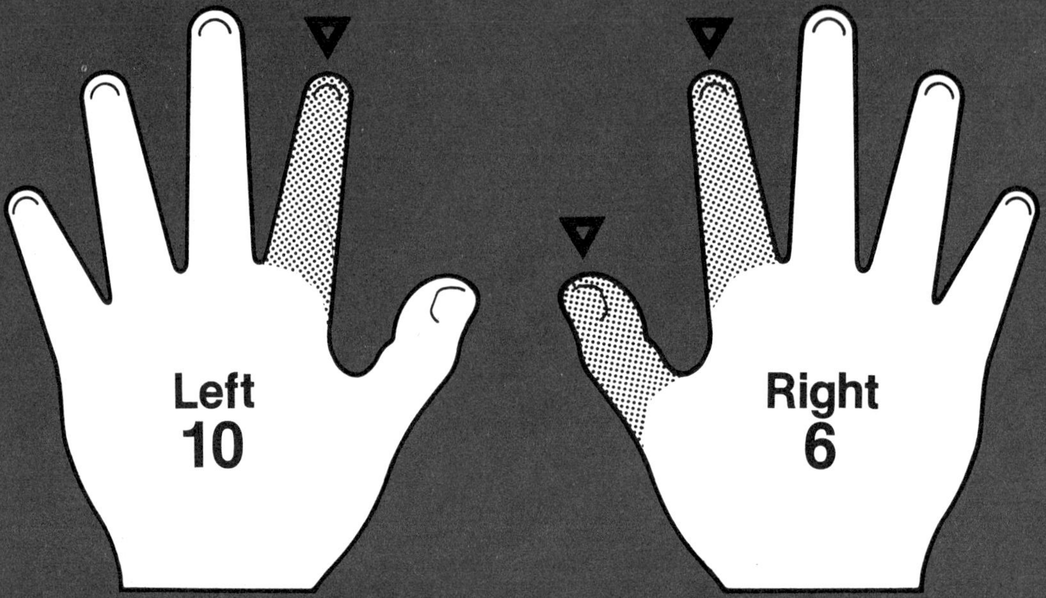

Left
10

Right
6

NUMBER **17**

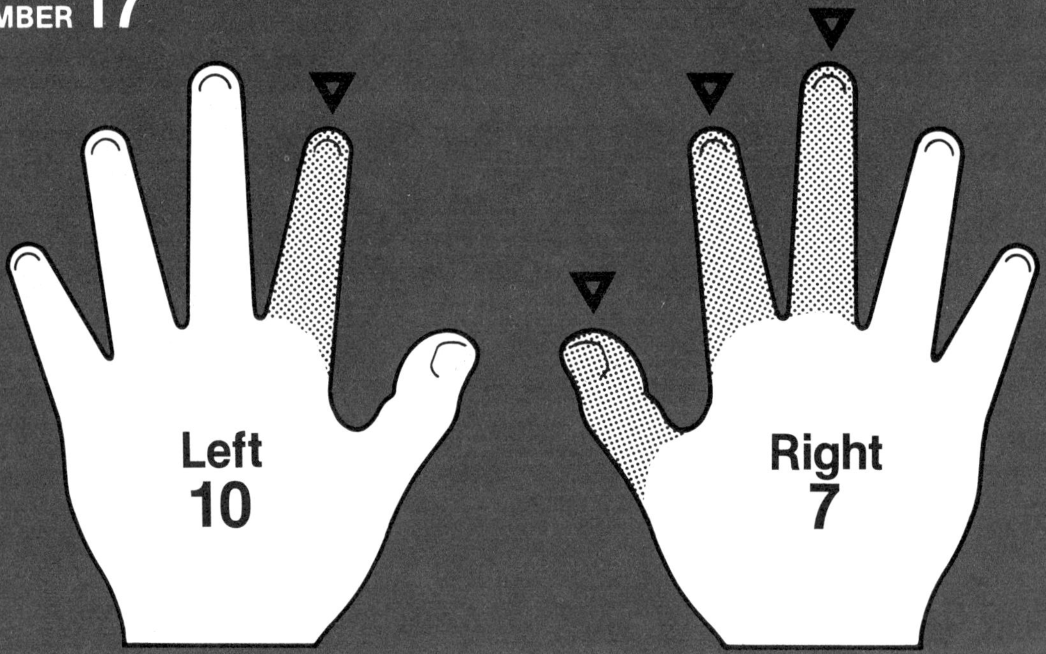

Left
10

Right
7

28

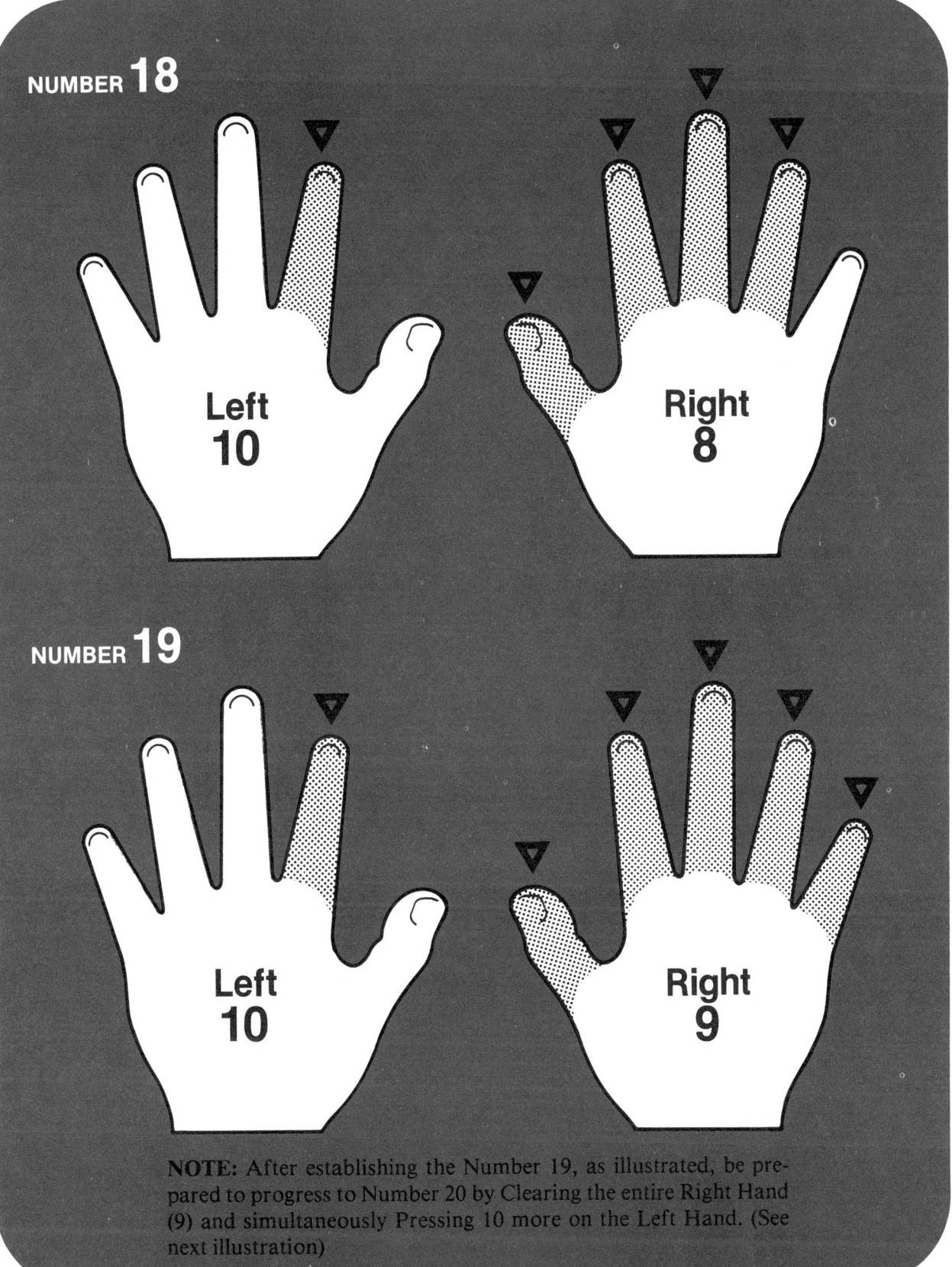

NUMBER 18

Left
10

Right
8

NUMBER 19

Left
10

Right
9

NOTE: After establishing the Number 19, as illustrated, be prepared to progress to Number 20 by Clearing the entire Right Hand (9) and simultaneously Pressing 10 more on the Left Hand. (See next illustration)

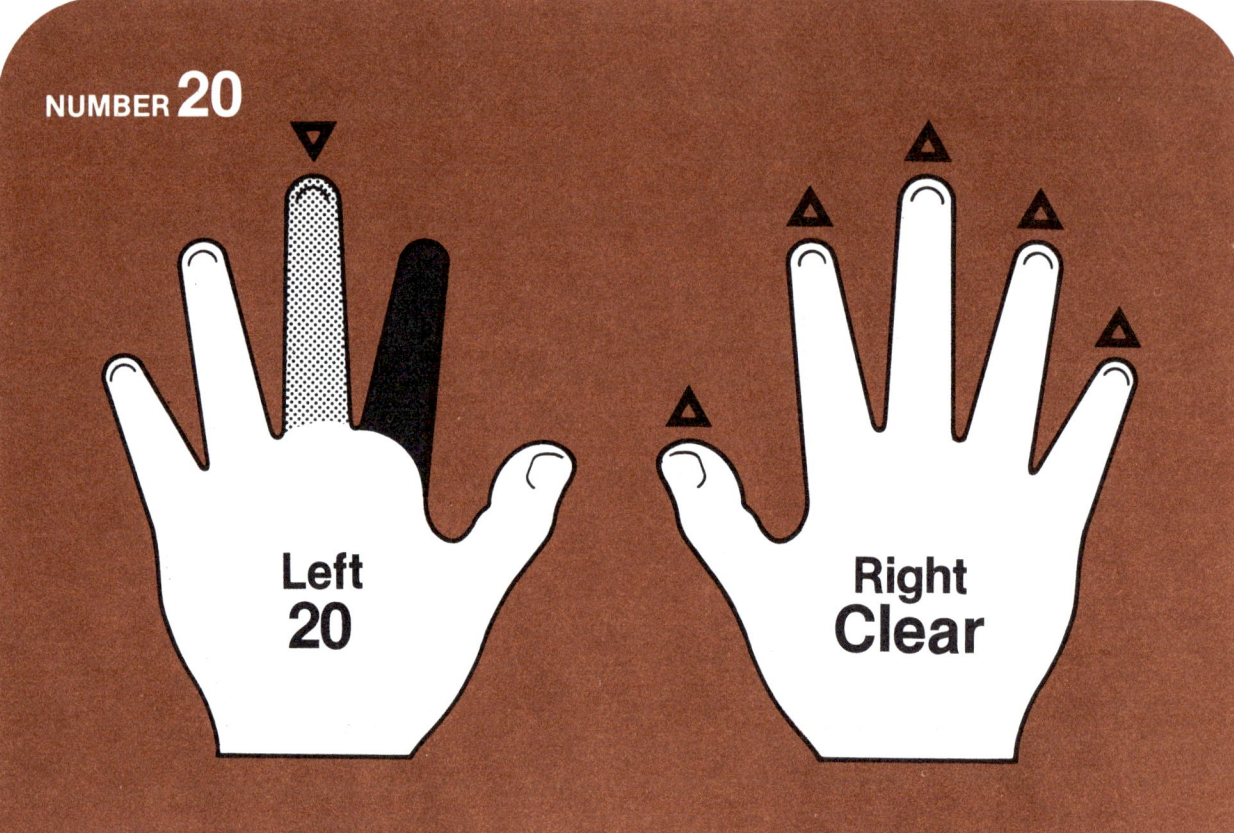

NUMBER 20

Left 20

Right Clear

NOTE: As previously explained, the consecutive count from 19 to 20 requires the Exchange of the Right Hand (9) for the Twentieth Finger of the Left Hand. Look again at the preceding diagram where the fingering for Number 19 is illustrated. To progress in our count to Number 20, we Clear the Right Hand, simultaneously Pressing the Twentieth Finger of the Left Hand.

After establishing 20, the Numbers from 21 to 29 are achieved by the Right Hand Press of required Fingers, while maintaining the Left Hand 20. In the same manner, use the Right Hand to achieve the Numbers:

31 to 39	71 to 79
41 to 49	81 to 89
51 to 59	91 to 99
61 to 69	

This diagram illustrates the progression from 29 to 30. Left Hand had previously registered 20 and Right Hand had previously registered 9. To achieve 30, simultaneously Press Thirtieth Finger on Left Hand and Clear the Right Hand.

NUMBER **30**

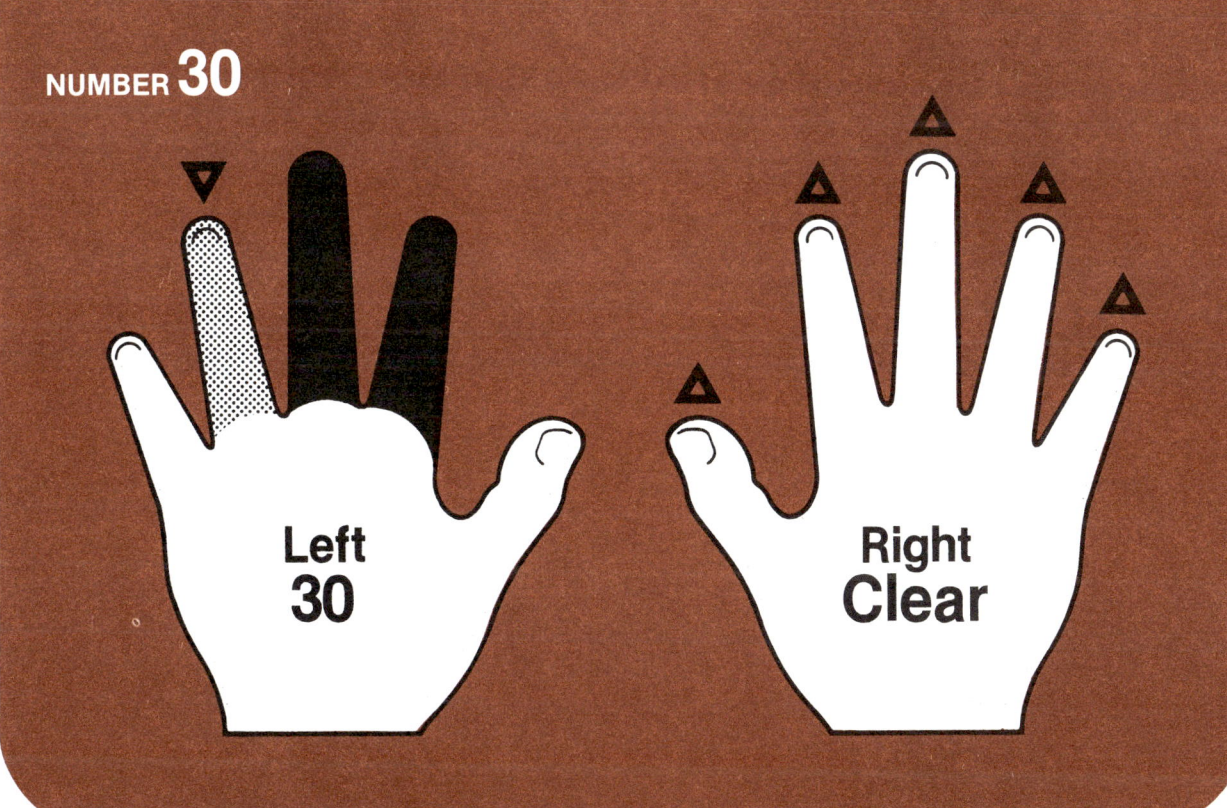

Left
30

Right
Clear

This diagram illustrates the progression from 39 to 40. Left Hand had previously registered 30 and Right Hand had previously registered 9. To achieve 40, simultaneously Press Fortieth Finger on Left Hand and Clear the Right Hand. One may continue progression to the Number 49 by now Pressing all fingers on Right Hand.

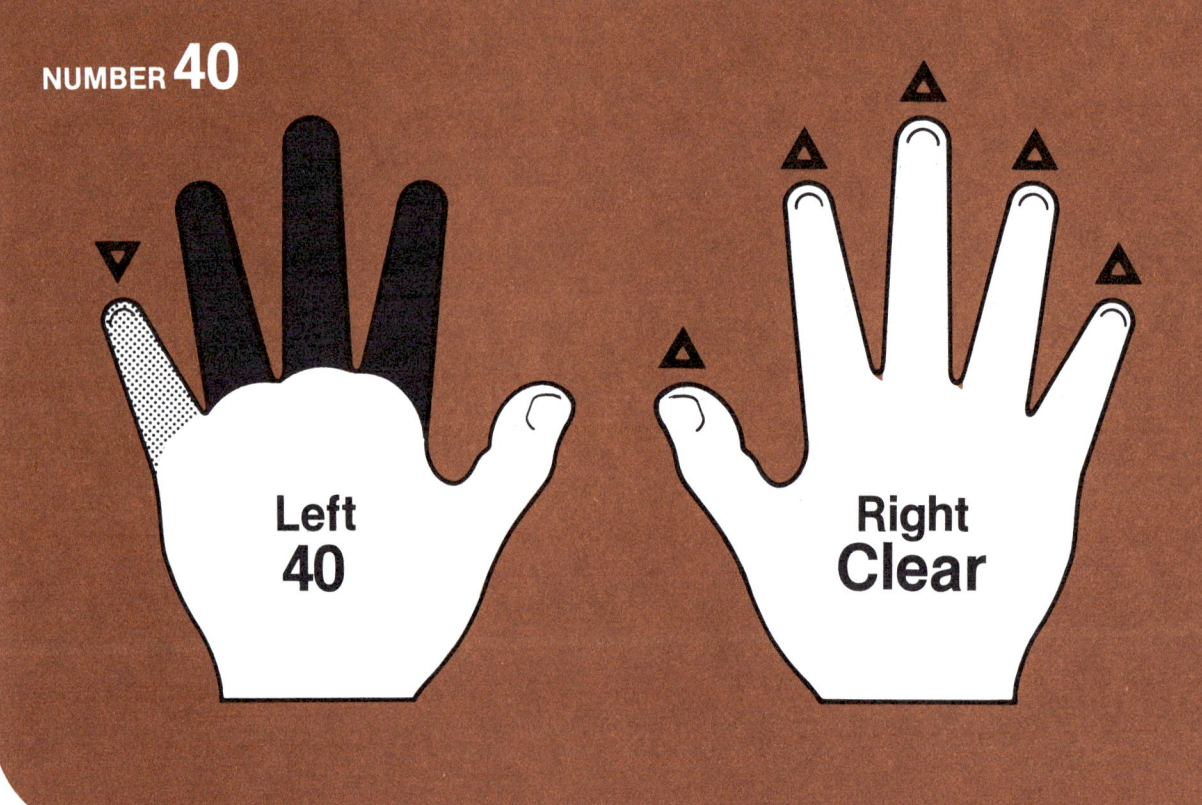

NUMBER **40**

Left
40

Right
Clear

**Left
50**

**Right
Clear**

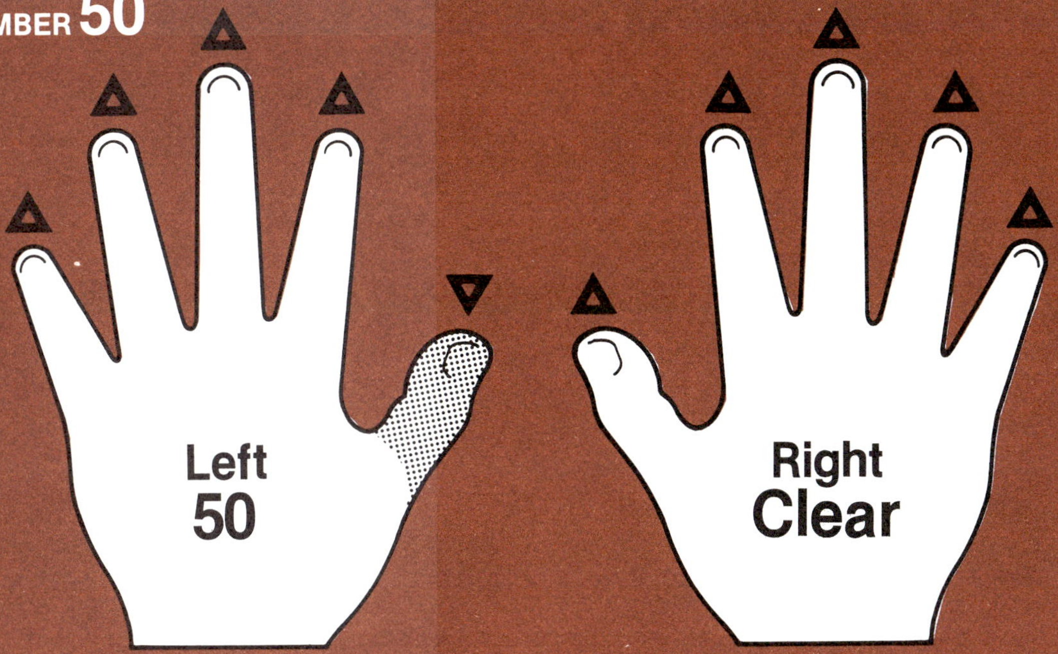

This diagram illustrates the progression from 49 to 50. At this juncture of consecutive numbers, the most far-reaching Exchange occurs. Left Hand had previously registered 40 and Right Hand had previously registered 9. To achieve 50, simultaneously Clear the Right Hand in Exchange for the next upcoming Tens Finger (Left Thumb) which, itself becomes the Exchange for the previously established Tenth, Twentieth, Thirtieth and Fortieth Fingers. A Double Exchange! Try it . . . it's simple.

First, Press 49.

Then, with a single manipulation, Press 50. Practice this Exchange several times, saying the numbers "Forty Nine" and "Fifty" out loud as you progress. Feel the ease of this transition.

This diagram illustrates the progression from 59 to 60. Left Hand had previously registered 50 and Right Hand had previously registered 9. To achieve 60, simultaneously Press Tenth Finger on Left Hand and Clear the Right Hand.

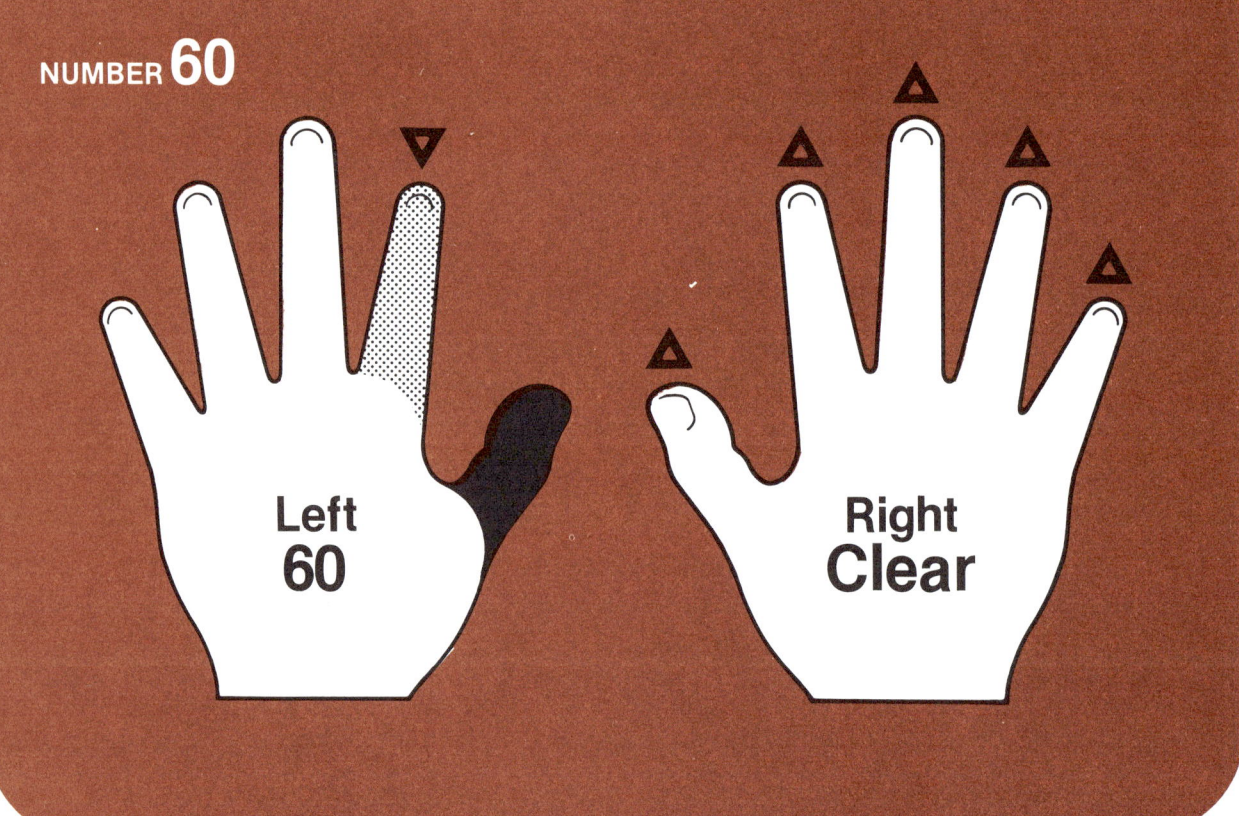

NUMBER **60**

Left
60

Right
Clear

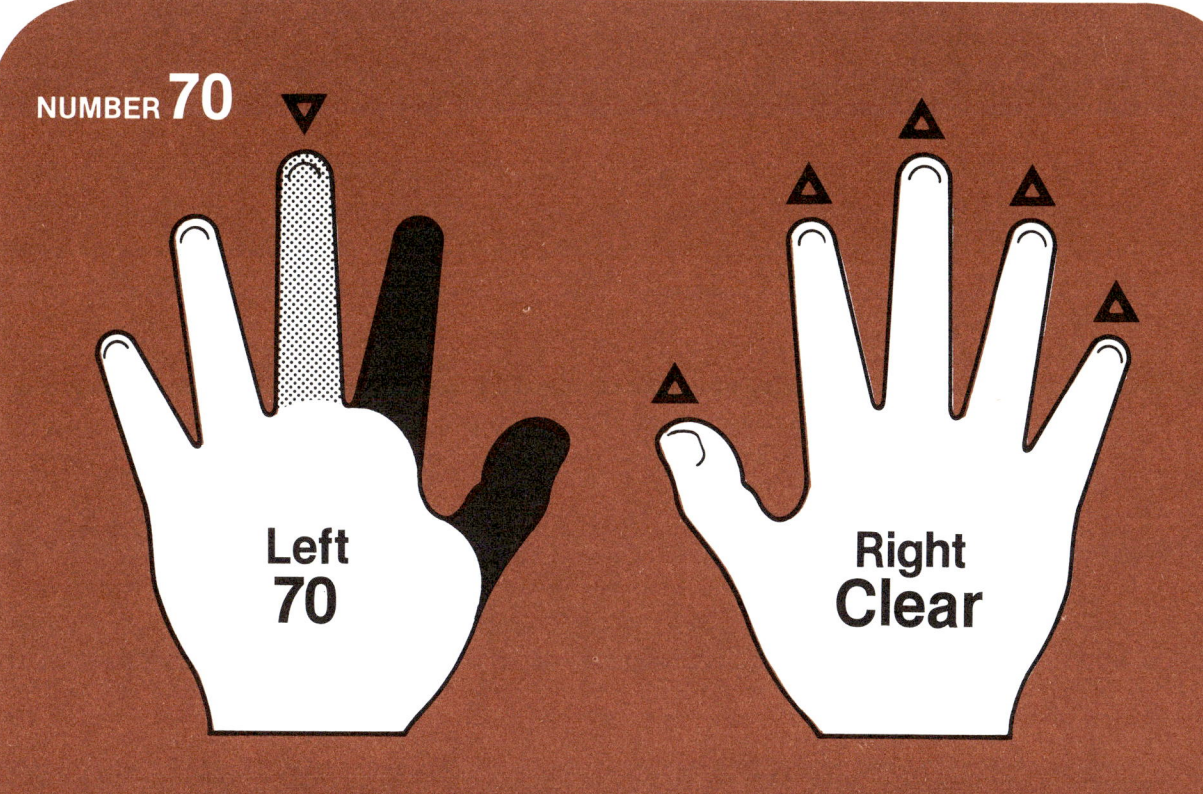

Left
70

Right
Clear

This diagram illustrates the progression from 69 to 70. Left Hand had previously registered 60 and Right Hand had previously registered 9. To achieve 70, simultaneously Press Twentieth Finger on Left Hand and Clear the Right Hand.

This diagram illustrates the progression from 79 to 80. Left Hand had previously registered 70 and Right Hand had previously registered 9. To achieve 80, simultaneously Press Thirtieth Finger on Left Hand and Clear the Right Hand.

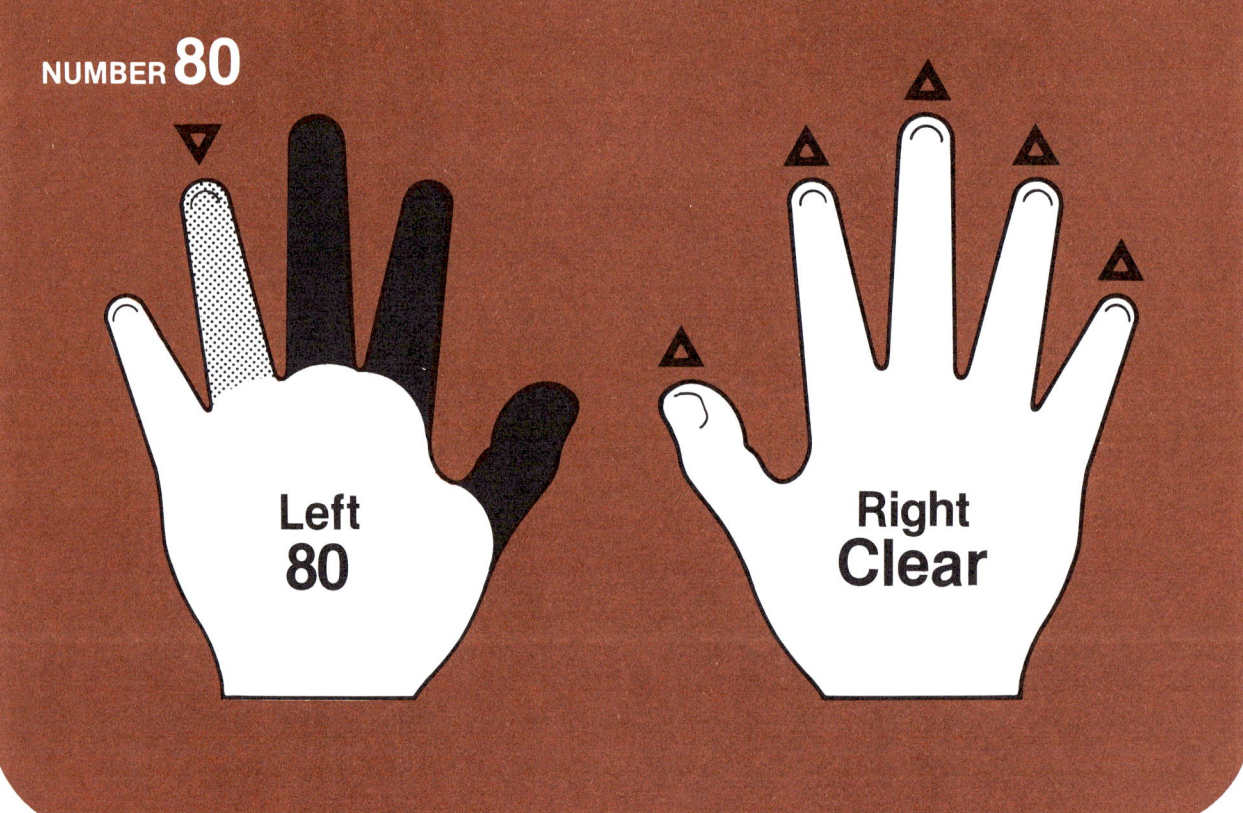

NUMBER **80**

Left
80

Right
Clear

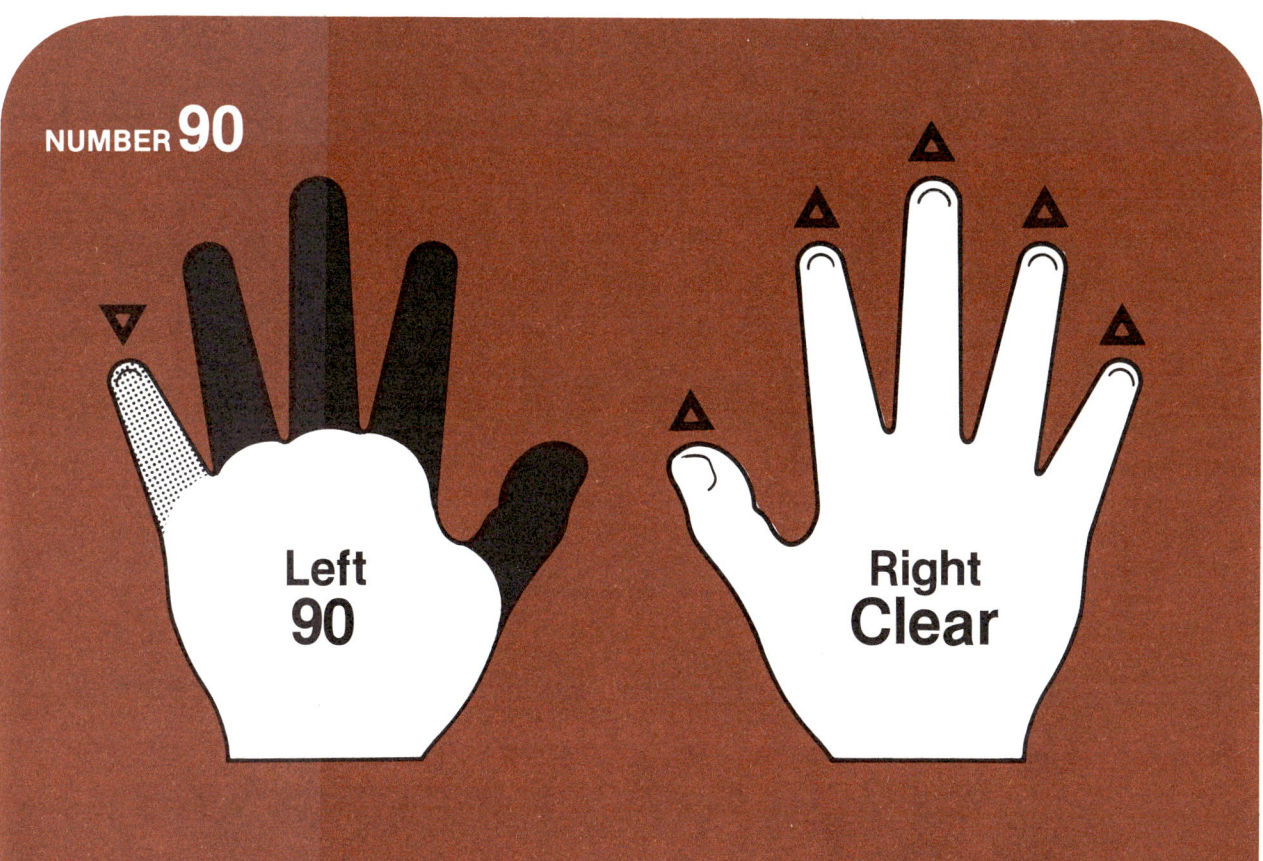

Left 90

Right Clear

This diagram illustrates the progression from 89 to 90. Left Hand had previously registered 80 and Right Hand had previously registered 9. To achieve 90, Press Fortieth Finger on Left Hand and Clear the Right Hand. Continue progression to the ultimate Number 99 by now Pressing all Fingers on the Right Hand.

How to Calculate

Having established a familiarity with our "tools" and an understanding of the values of each finger, plus a knowledge of the progression from Number 1 to Number 99, we are prepared to use this information to perform calculations of Addition.

There are several groups of single digit numbers which the reader must learn to combine. They are treated here by their increasing degree of difficulty . . . from the easiest to the hardest.

$$1 + 1 = \square$$

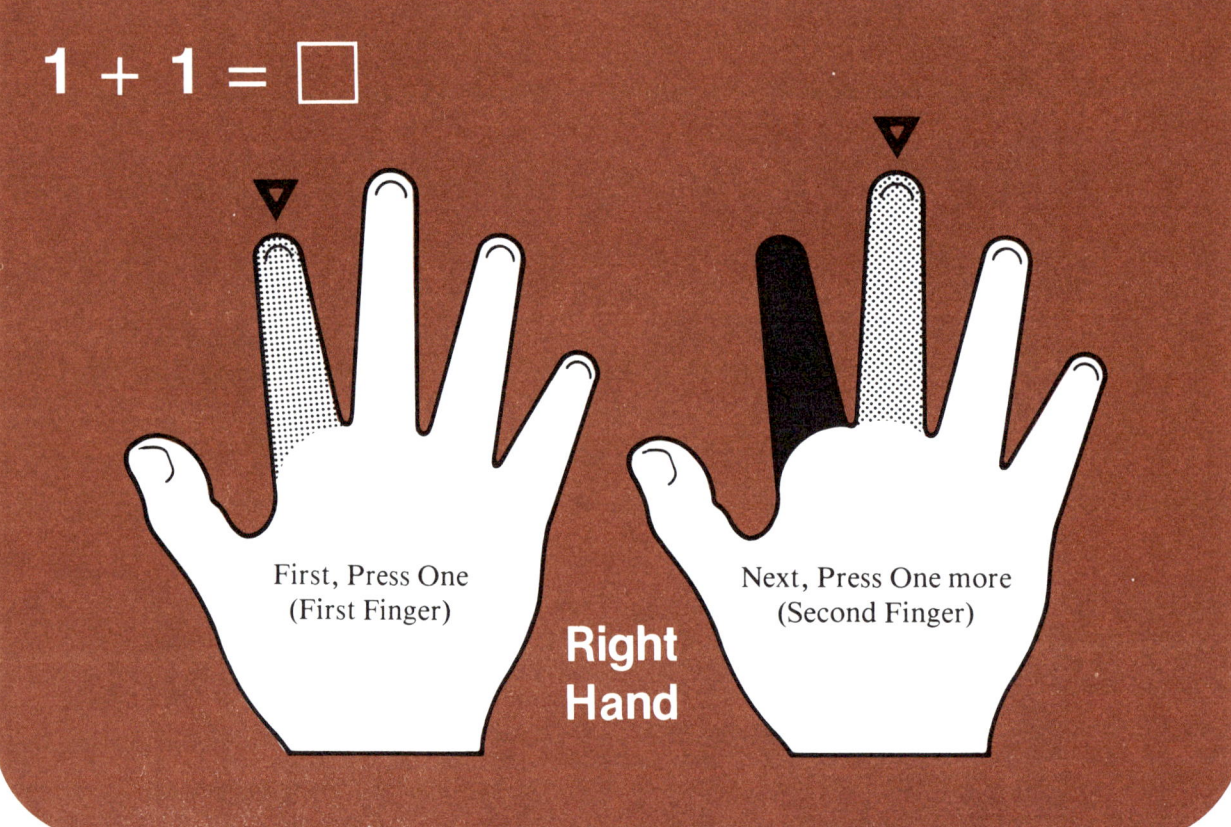

First, Press One
(First Finger)

Next, Press One more
(Second Finger)

Right Hand

The proper dialogue between parent and child for this calculation will be:

We are going to add two numbers together. Let's see what happens. Watch me first. I am going to add One plus One more. (Write 1 + 1 = □) **First, I will Press One.**

Then, I will Press One more.

Now I have Two. (Clear your fingers) **Now let's do it together. Ready? Press One.**

Child: (Presses First Finger)
And Press One more.
Child: (Presses Second Finger)
How Many?
Child: Two!
Very good. Now . . . let's go on. First, Clear your fingers.
Child: (Clears fingers) (Write 1 + 2 = □)

1 + 2 = ☐

PRESS ONE (First Finger)   PRESS TWO MORE (Second & Third Fingers)

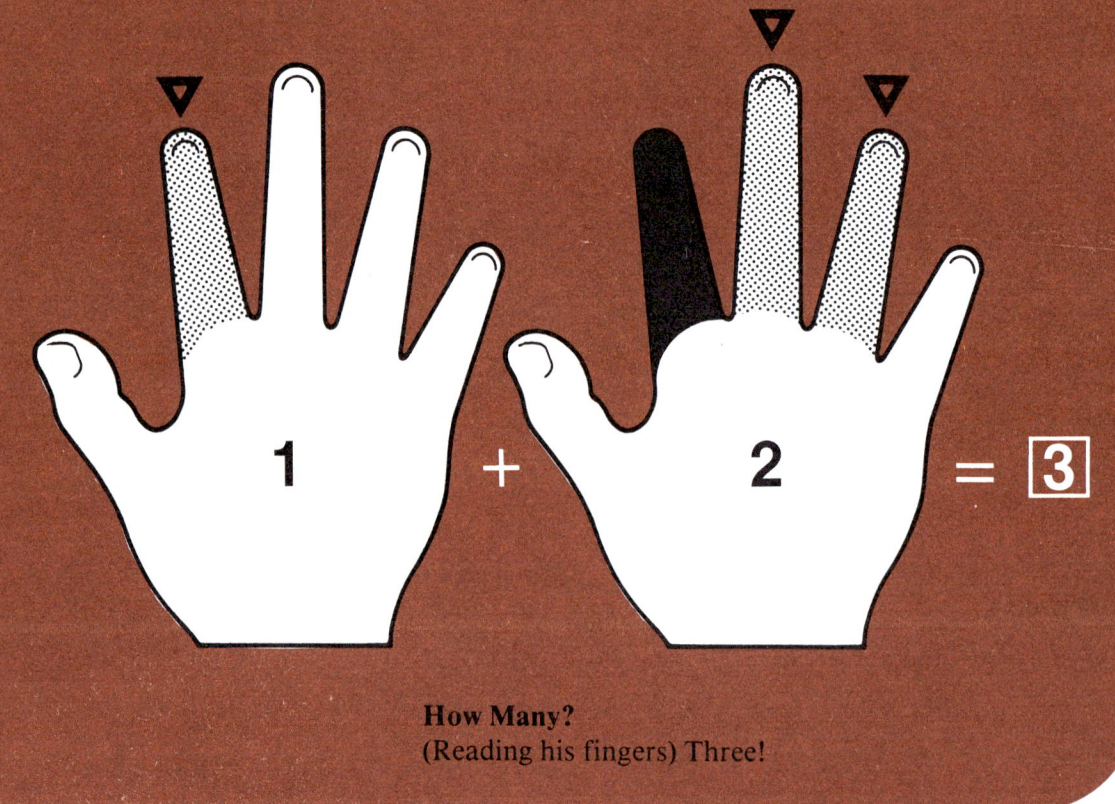

1 + 2 = 3

How Many?
(Reading his fingers) Three!

Continue to demonstrate and then have child repeat the following:

1 + 3 = □

PRESS ONE (First Finger)   **PRESS THREE MORE** (Second, Third & Fourth Fingers)

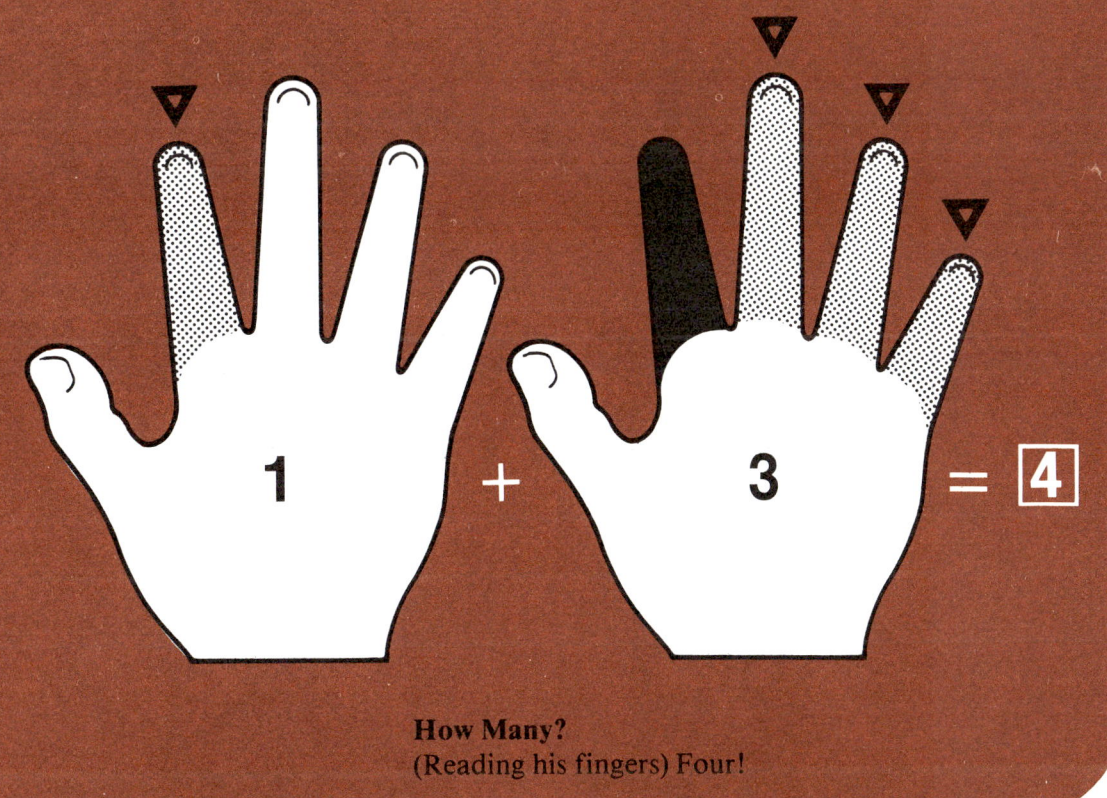

How Many?
(Reading his fingers) Four!

1 + 5 = ☐

PRESS ONE (First Finger)   **PRESS FIVE MORE** (Fifth Finger)

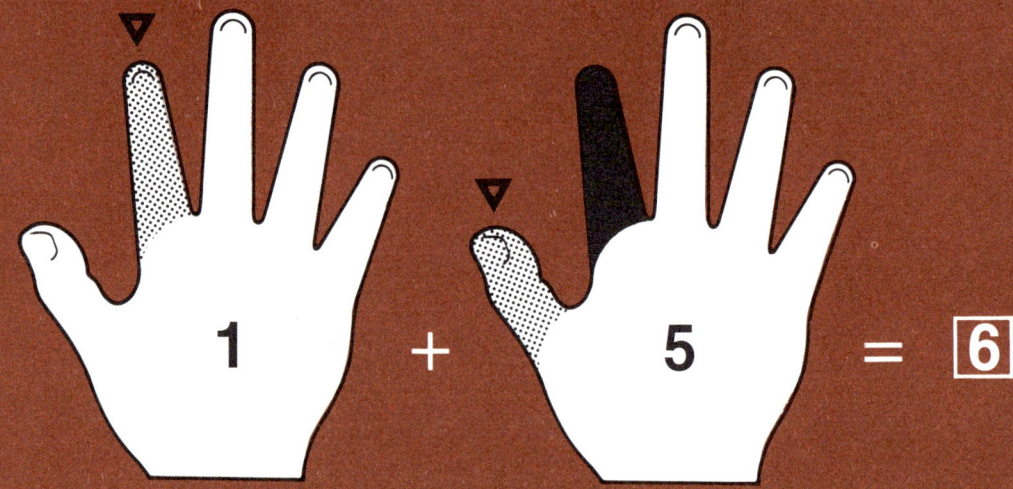

NOTE: Once again, as explained earlier, the First Stage child will not be capable of comprehending this calculation as a single step from 1 to 6 by simply Pressing the Fifth Finger. Such Addition must first be approached by counting each single unit in sequence to reach 6. In this example, instruct:

    Press One,
    Now Press Five more.

The child having Pressed the First Finger (1) will then Press the Second Finger, beginning his *new* count, saying: "One" then "Two, Three, Four, Five", Pressing the other fingers in sequence and ending with his hand as shown at left.

How Many?

Child: (Reading his fingers) Six!

It is evident again that each finger, *when used in sequence,* is only 1. Therefore, if one wishes to add 5 to any existing number, he begins his *added* count with "One", knowing when he reaches "Five" his fingers can be "read" for the result. In the above example, when adding 5 to 1 and counting orally in sequence, as described, the *word* "Four" will be reached as the Thumb is pressed—even though the cumulative sum at that point will be 5. This is further evidence of the value of 1 represented by each finger.

In all of the following exercises, the above single unit (First Stage) approach must be taken initially. Later, all examples must be done again, repeatedly, by using the more advanced (Second and Third Stages) methods, as illustrated.

Regardless of the Stage of progress, the example being addressed must always be written and the child must copy example on his own paper, stating it orally.

The Four Stages of Manipulation
$(1 + 6 = \square)$

First Stage: Count each single unit in sequence. Having Pressed the First Finger (1), pupil will then Press the Second Finger, beginning his *new* count, saying: "One" then "Two, Three, Four, Five, Six," Pressing the other fingers in sequence and ending with his hand as shown at right.
How Many?

Child: (Reading his fingers) Seven!

Second Stage: Press First Finger (1). Then Press Fifth Finger (5 more). Then Press Second Finger (1 more).
Press One.

Child: (Pressing First Finger) One
Now we must add 6. First Press 5.

Child: (Pressing Fifth Finger) Five.
And One More.

Child: (Pressing Second Finger) Six.
How Many?

Child: (Reading his fingers) Seven!

Third Stage: Press First Finger (1). Then Press Fifth Finger and Second Finger simultaneously (6)
Press One.

Child: (Pressing First Finger) One.
Now Press 6 more.
(Pupil Presses Fifth Finger and Second Finger simultaneously)
How Many?

Child: (Reading his fingers) Seven!

Fourth Stage: (This is the ultimate achievement. To reach this Stage one must have practiced each combination of numbers so frequently that there is an instant recognition of the total, which then is Pressed as a single manipulation. The CHISANBOP Workbook deals with these combinations according to their increasing degree of difficulty.)

Pupil recognizes $1 + 6$ as the embodiment of 7 and simultaneously Presses Fifth, First and Second Fingers for the total. This Stage becomes significant when adding a column of numbers. Then, instead of Pressing each number individually, one embraces two numbers—or even three—simultaneously, bypassing number-by-number addition.

1 + 6 = ☐

NOTE: The calculations illustrated on this page and on succeeding pages all show Second and Third Stage manipulations. However, the beginner is to avoid these advanced Stages and use First Stage (Unit-by-Unit) counting *only*. To skip the First Stage—even if there is an apparent understanding of the advanced stages—would be a serious choice and one which will adversely affect later manipulations.

PRESS ONE (First Finger)

PRESS SIX MORE (Fifth & Second Fingers)
In the First Stage, one would add 6 unit-by-unit.

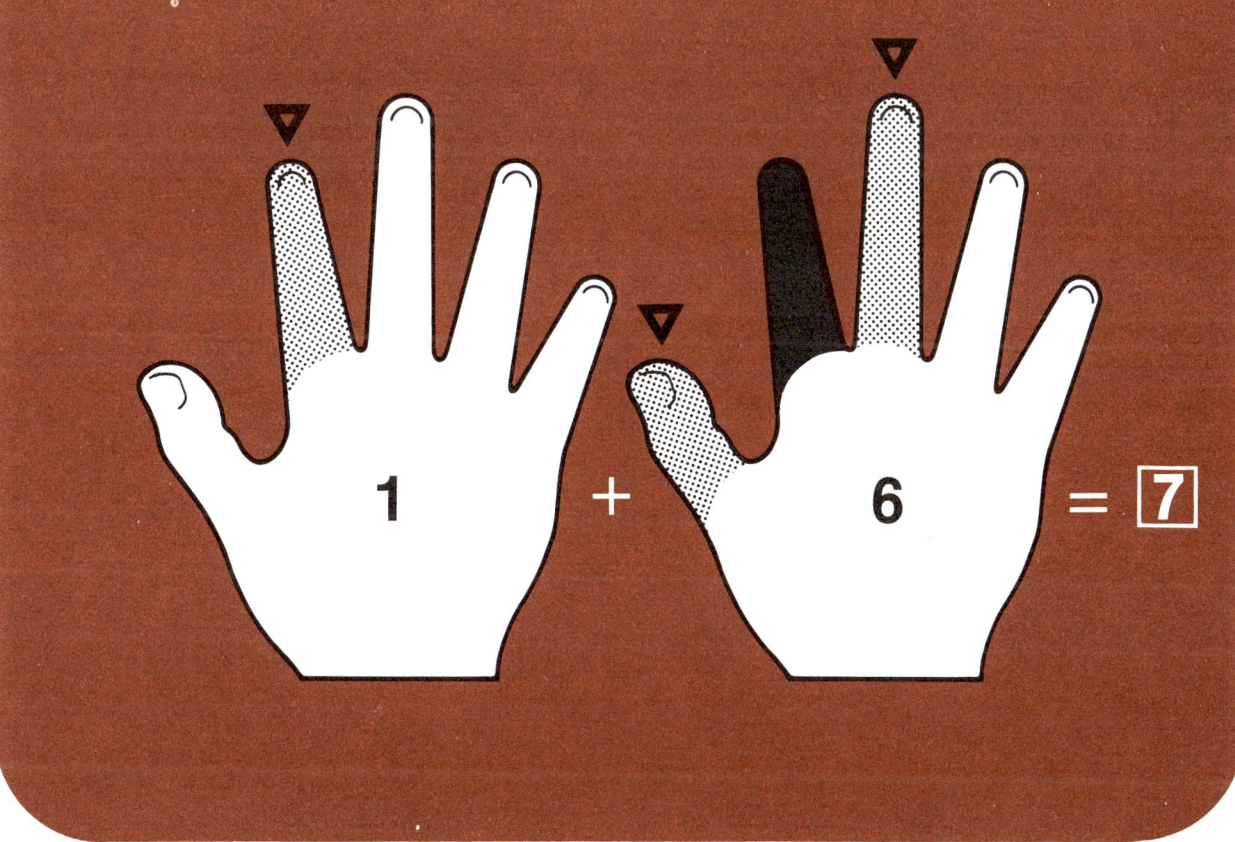

1 + 7 = ☐

PRESS ONE (First Finger)

PRESS SEVEN MORE (Fifth, Second and Third Fingers)

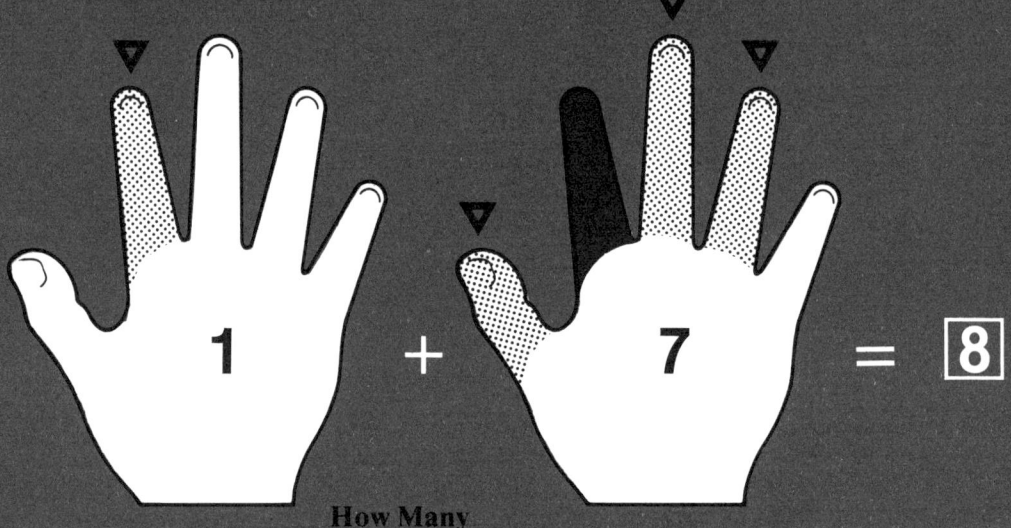

How Many
(Reading his fingers) Eight!

1 + 8 = ☐

PRESS ONE (First Finger)

PRESS EIGHT MORE (Fifth, Second, Third and Fourth Fingers)

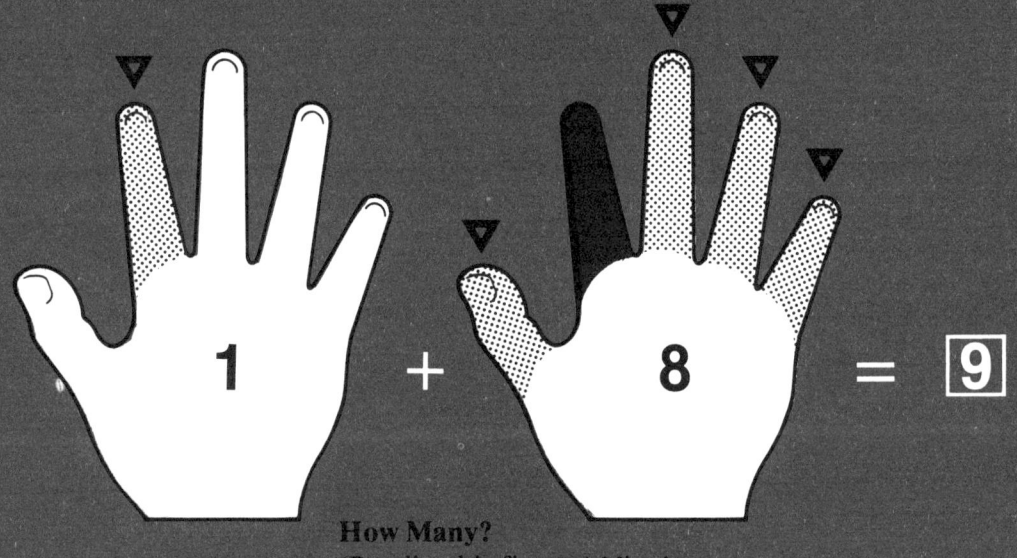

How Many?
(Reading his fingers) Nine!

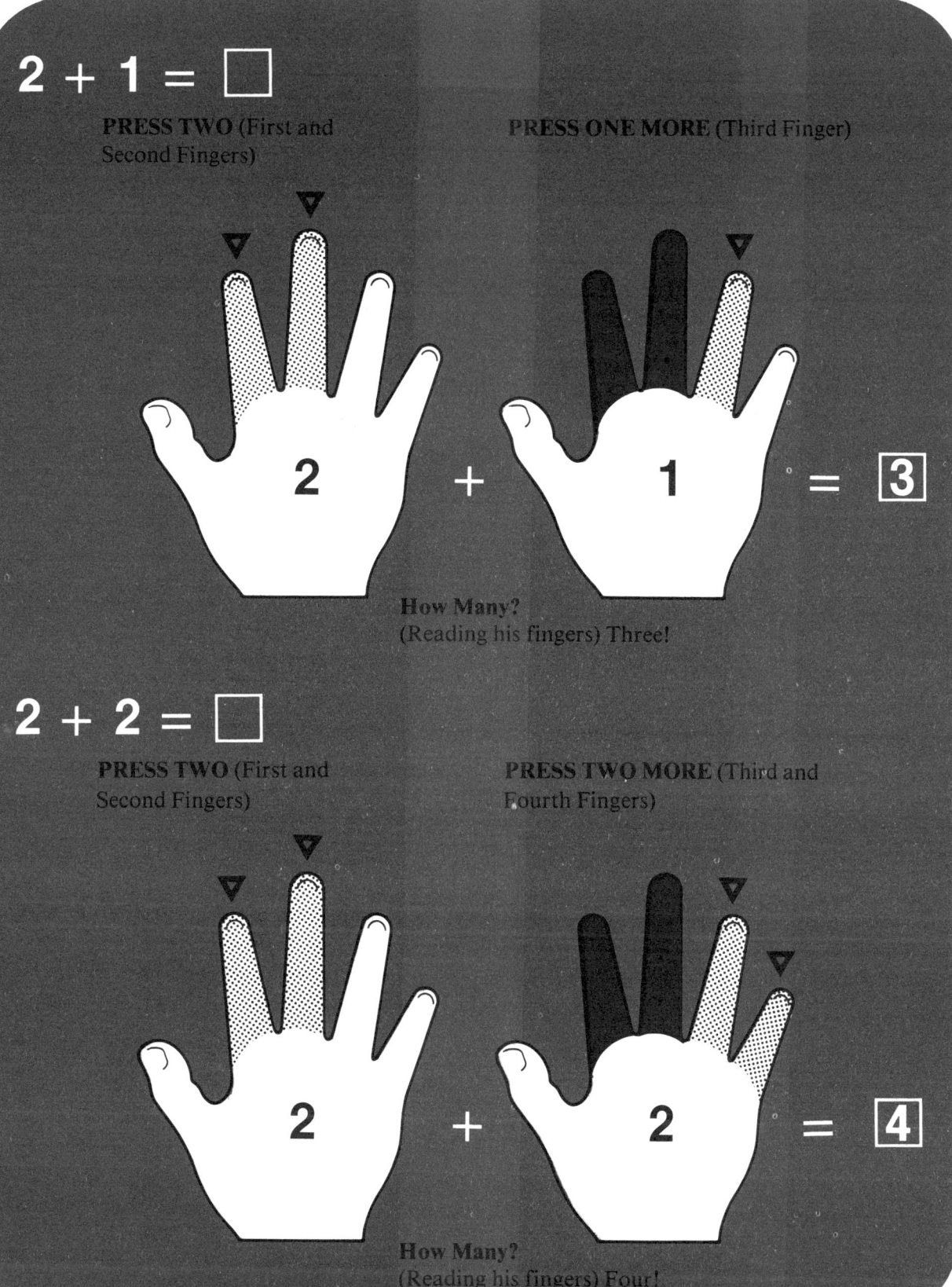

2 + 1 = ☐

PRESS TWO (First and Second Fingers)

PRESS ONE MORE (Third Finger)

2 + 1 = ③

How Many?
(Reading his fingers) Three!

2 + 2 = ☐

PRESS TWO (First and Second Fingers)

PRESS TWO MORE (Third and Fourth Fingers)

2 + 2 = ④

How Many?
(Reading his fingers) Four!

2 + 5 = □

PRESS TWO (First and Second Fingers)

PRESS FIVE MORE (Fifth Finger)

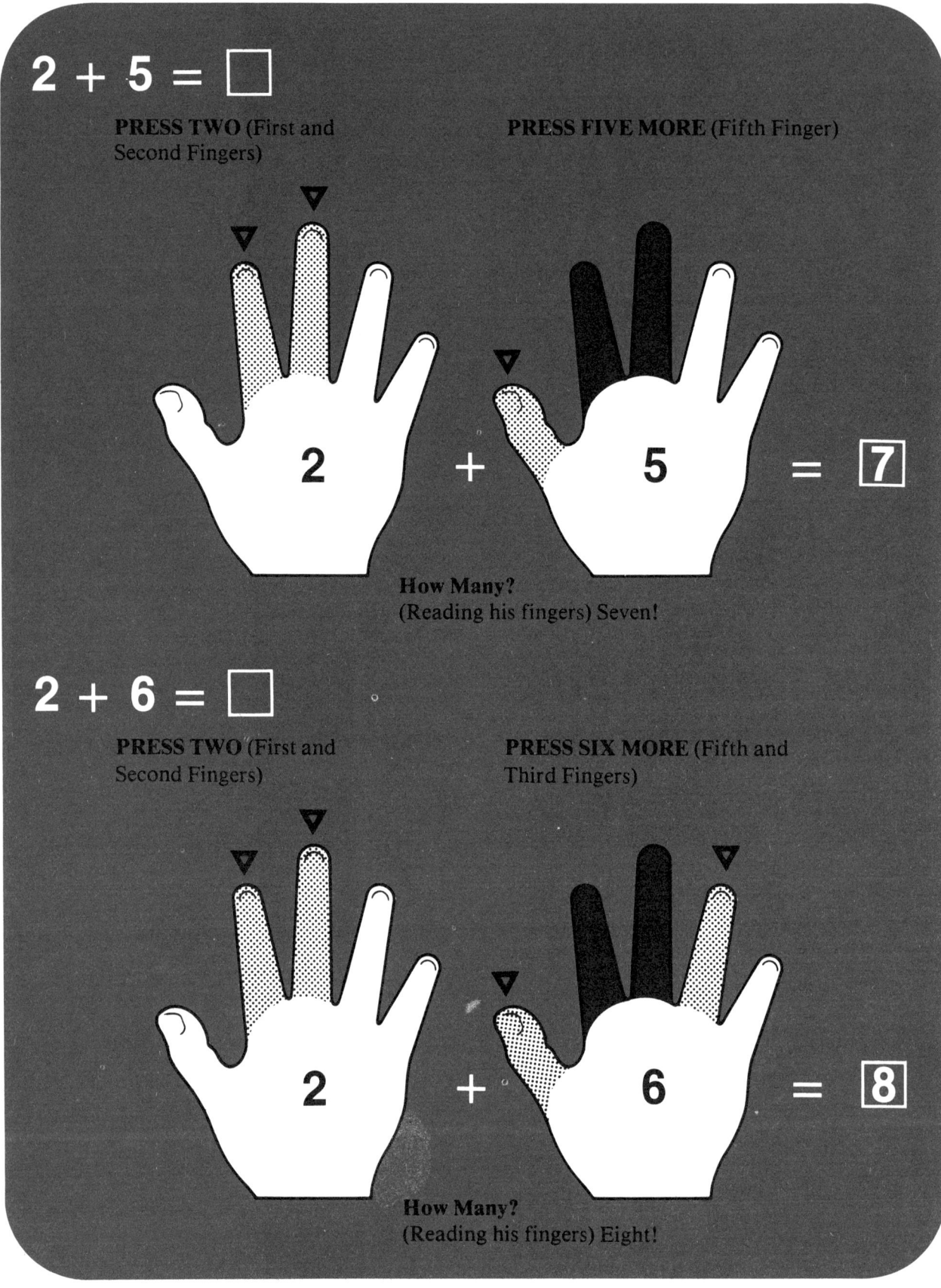

2 + = 5 = 7

How Many?
(Reading his fingers) Seven!

2 + 6 = □

PRESS TWO (First and Second Fingers)

PRESS SIX MORE (Fifth and Third Fingers)

2 + 6 = 8

How Many?
(Reading his fingers) Eight!

2 + 7 = ☐

PRESS TWO (First and Second Fingers)

PRESS SEVEN MORE (Fifth, Third and Fourth Fingers)

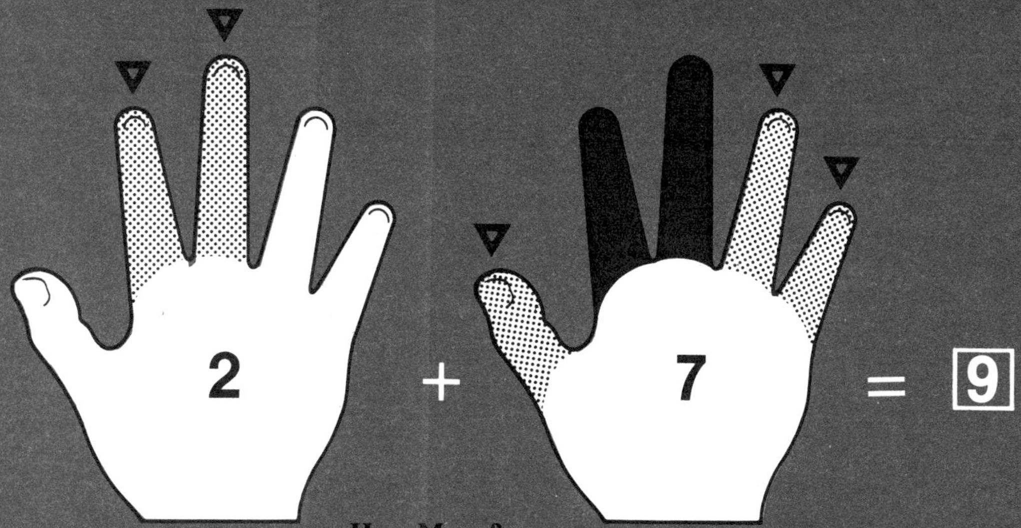

2 + **7** = **9**

How Many?
(Reading his fingers) Nine!

3 + 1 = ☐

PRESS THREE (First, Second and Third Fingers)

PRESS ONE MORE (Fourth Finger)

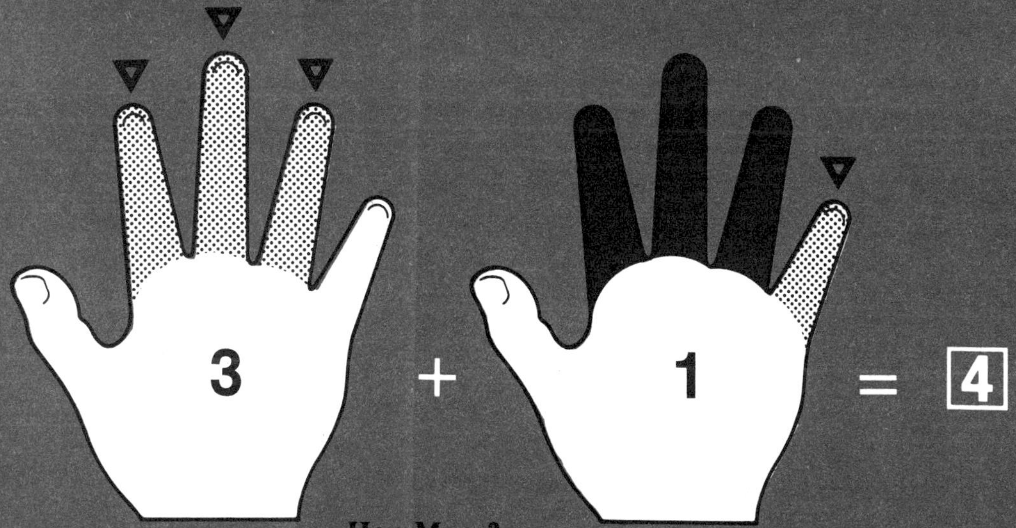

3 + **1** = **4**

How Many?
(Reading his fingers) Four!

3 + 5 = ☐

PRESS THREE (First, Second and Third Fingers)　　　　**PRESS FIVE MORE** (Fifth Finger)

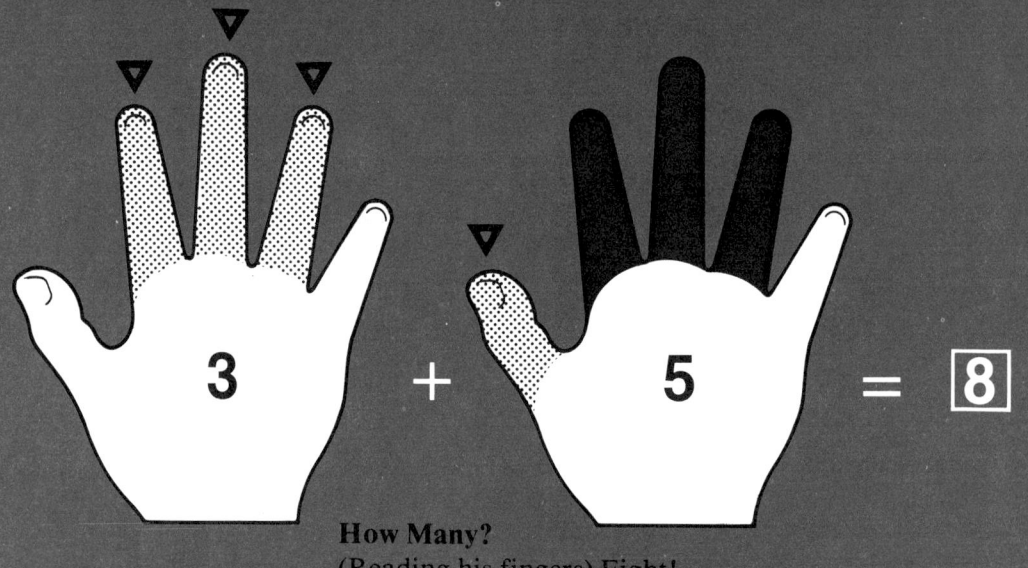

3 + 5 = 8

How Many?
(Reading his fingers) Eight!

3 + 6 = ☐

PRESS THREE (First, Second and Third Fingers)　　　　**PRESS SIX MORE** (Fifth and Fourth Fingers)

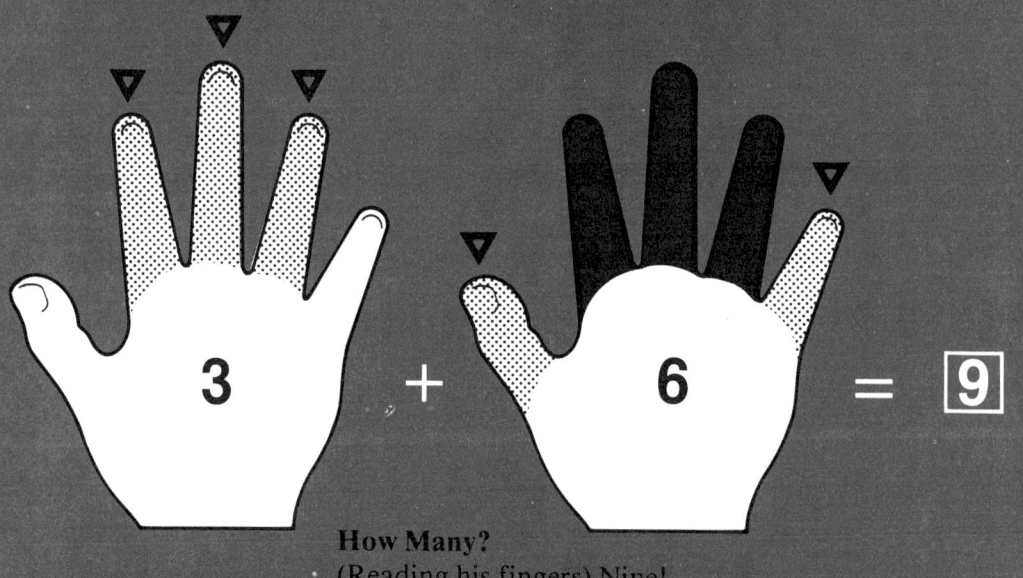

3 + 6 = 9

How Many?
(Reading his fingers) Nine!

4 + 5 = ☐

PRESS FOUR (First, Second, Third and Fourth Fingers)

PRESS FIVE MORE (Fifth Finger)

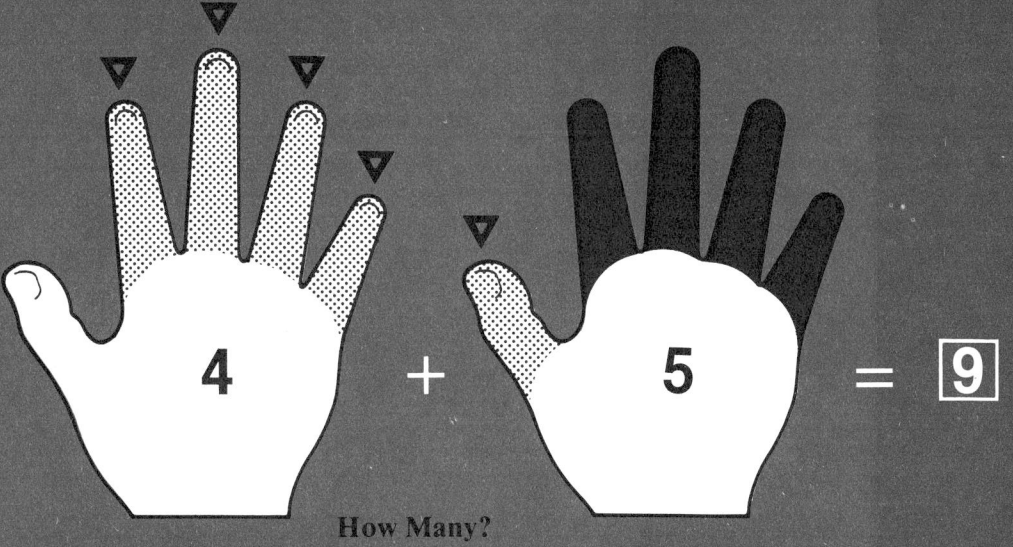

4 + **5** = 9

How Many?
(Reading his fingers) Nine!

5 + 1 = ☐

PRESS FIVE (Fifth Finger)

PRESS ONE MORE (First Finger)

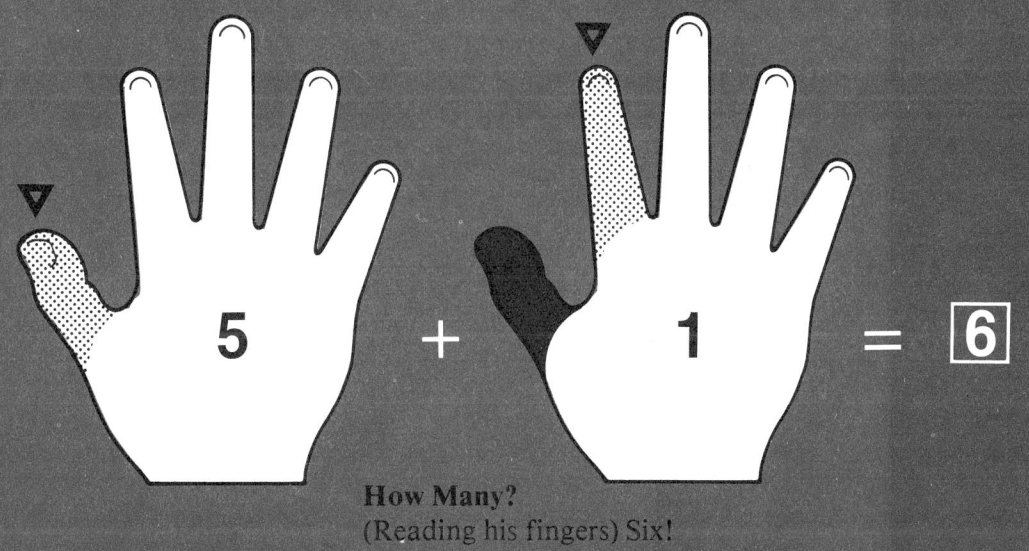

5 + **1** = 6

How Many?
(Reading his fingers) Six!

5 + 2 = ☐

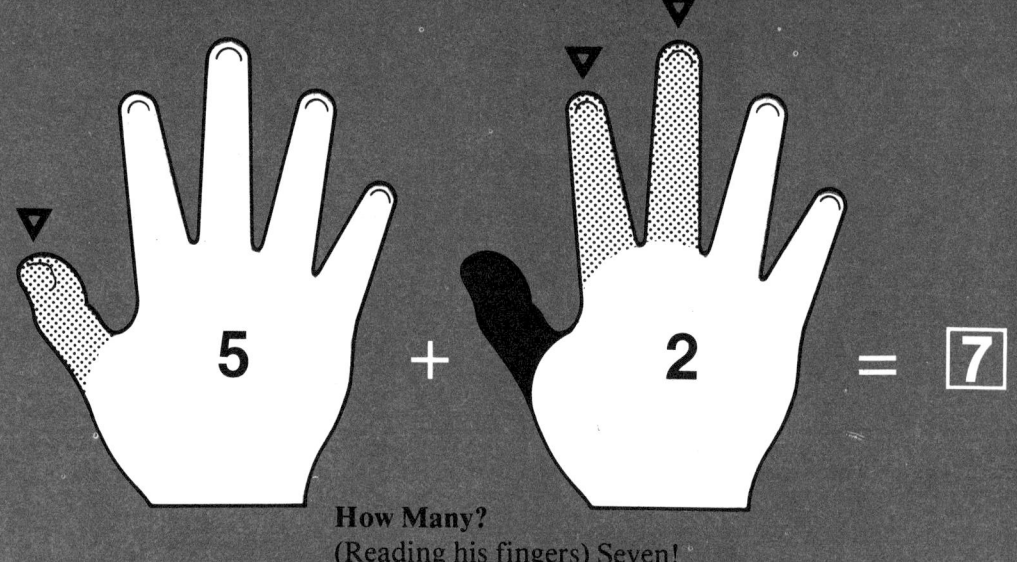

PRESS FIVE (Fifth Finger)

PRESS TWO MORE (First and Second Finger)

5 + 2 = 7

How Many?
(Reading his fingers) Seven!

5 + 3 = ☐

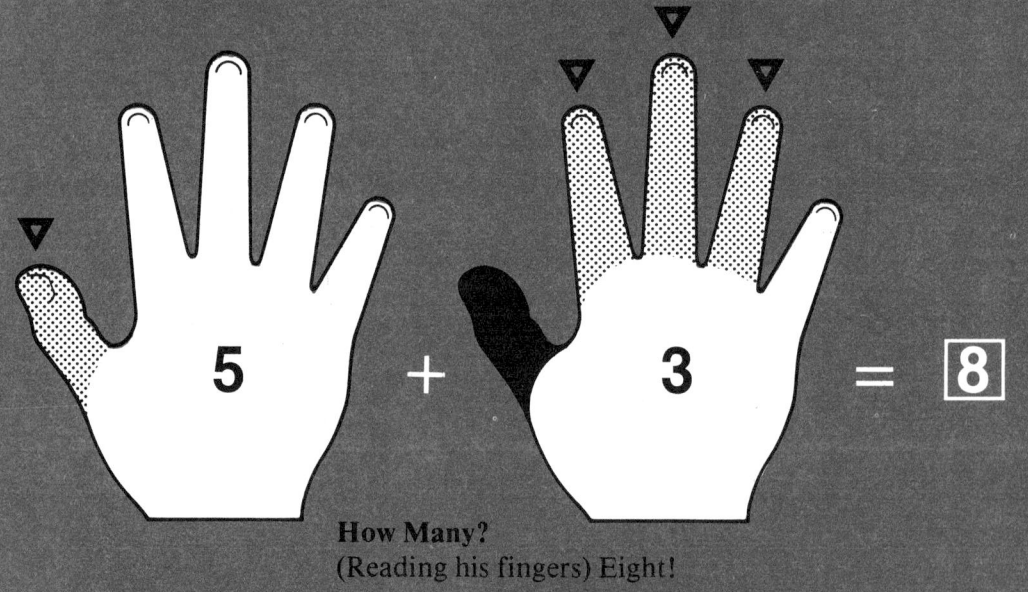

PRESS FIVE (Fifth Finger)

PRESS THREE MORE (First, Second and Third Fingers)

5 + 3 = 8

How Many?
(Reading his fingers) Eight!

5 + 4 = ☐

PRESS FIVE (Fifth Finger)

PRESS FOUR MORE (First, Second, Third and Fourth Fingers)

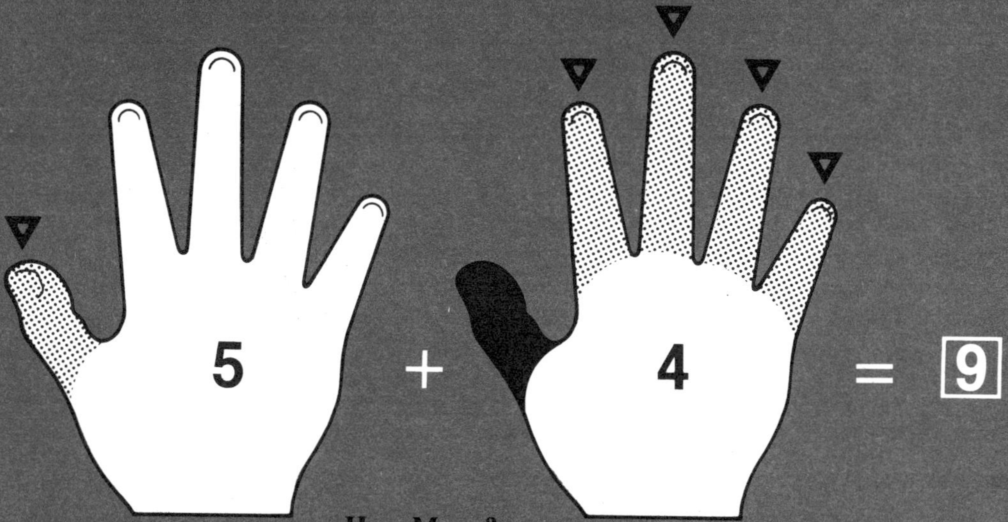

5 + 4 = 9

How Many?
(Reading his fingers) Nine!

6 + 1 = ☐

PRESS SIX (Fifth and First Fingers)

PRESS ONE MORE (Second Finger)

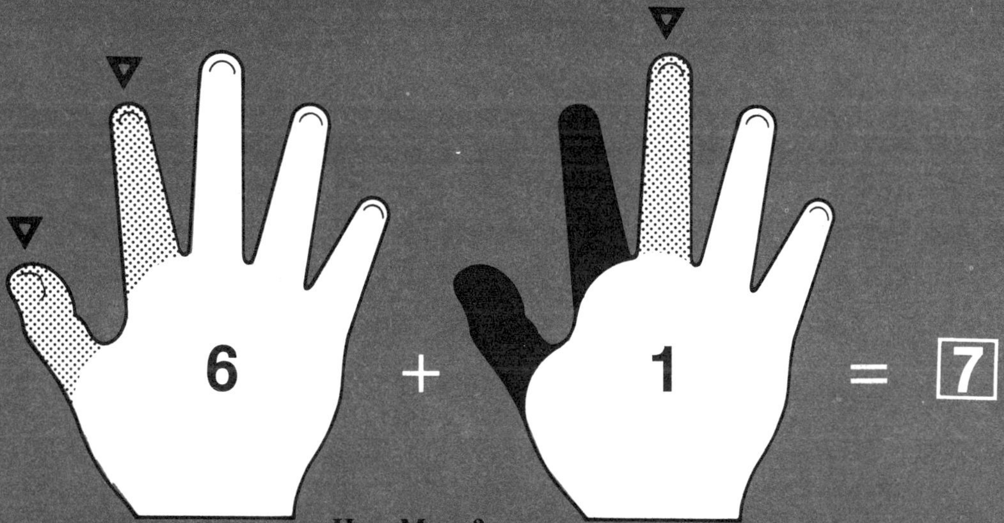

6 + 1 = 7

How Many?
(Reading his fingers) Seven!

6 + 2 = ☐

PRESS SIX (Fifth and First Fingers)

PRESS TWO MORE (Second and Third Fingers)

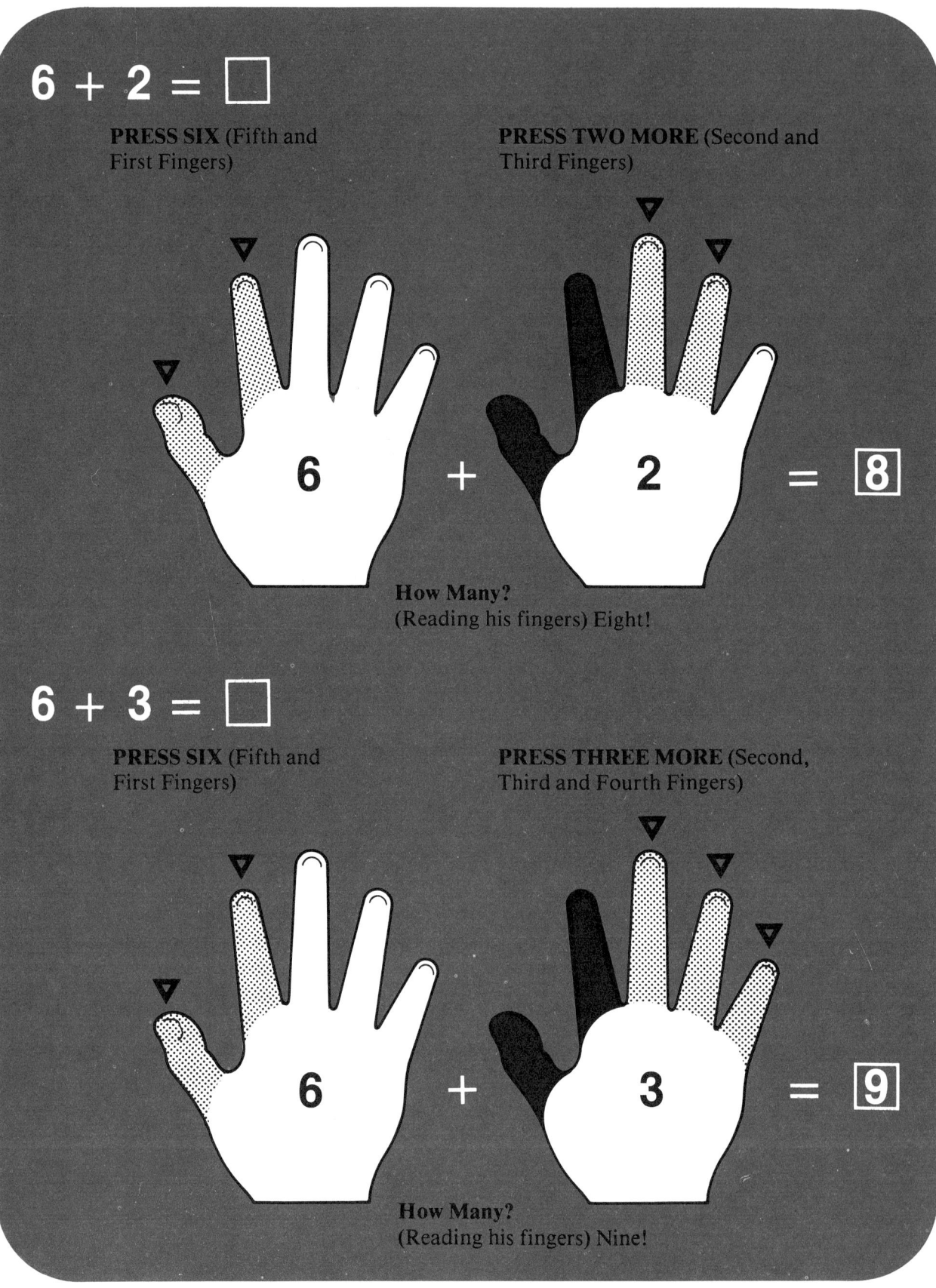

6 + 2 = 8

How Many?
(Reading his fingers) Eight!

6 + 3 = ☐

PRESS SIX (Fifth and First Fingers)

PRESS THREE MORE (Second, Third and Fourth Fingers)

6 + 3 = 9

How Many?
(Reading his fingers) Nine!

7 + 1 = ☐

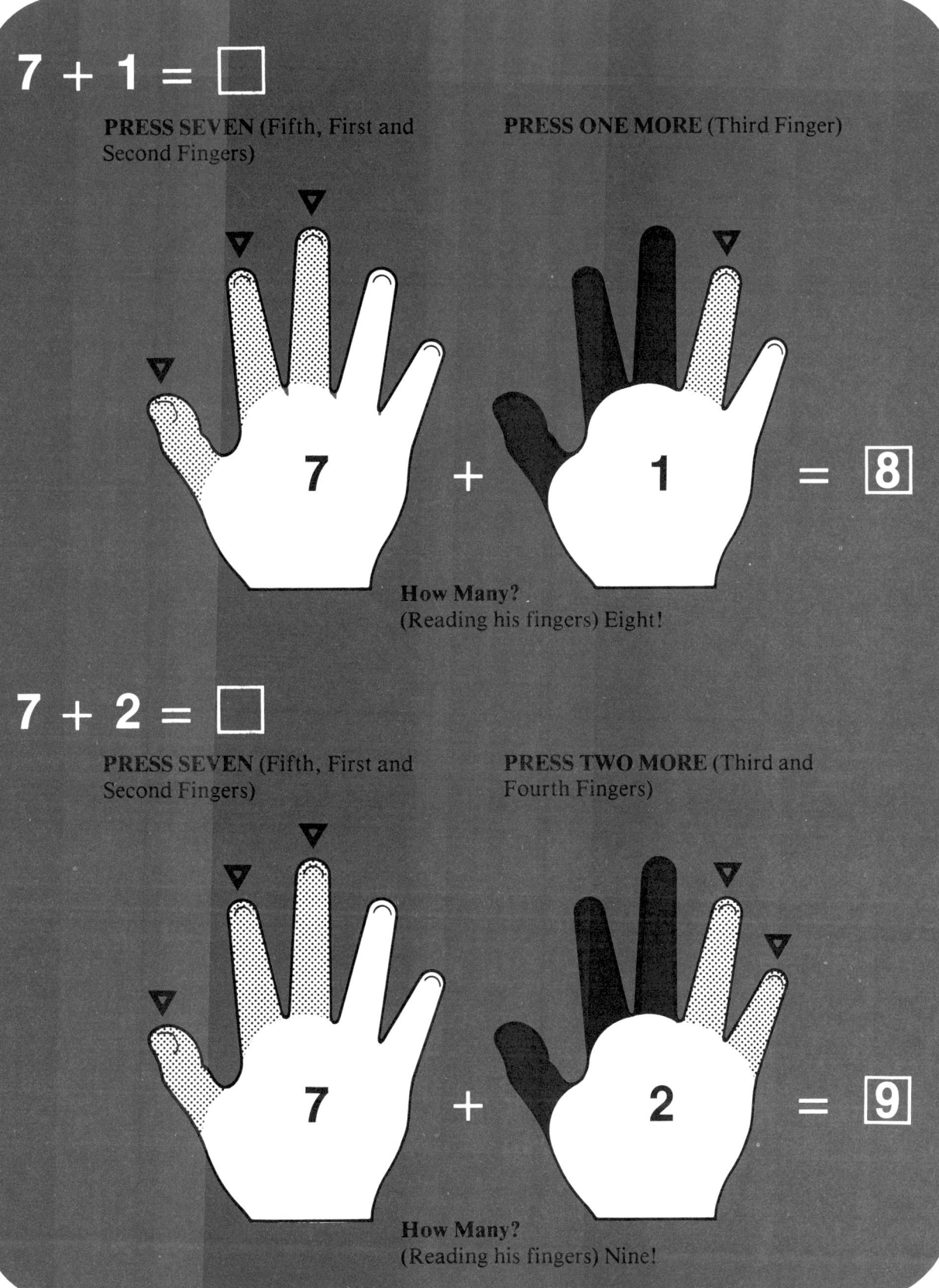

PRESS SEVEN (Fifth, First and Second Fingers)

PRESS ONE MORE (Third Finger)

7 + 1 = 8

How Many?
(Reading his fingers) Eight!

7 + 2 = ☐

PRESS SEVEN (Fifth, First and Second Fingers)

PRESS TWO MORE (Third and Fourth Fingers)

7 + 2 = 9

How Many?
(Reading his fingers) Nine!

8 + 1 = ☐

PRESS EIGHT (Fifth, First, Second and Third Fingers)

PRESS ONE MORE (Fourth Finger)

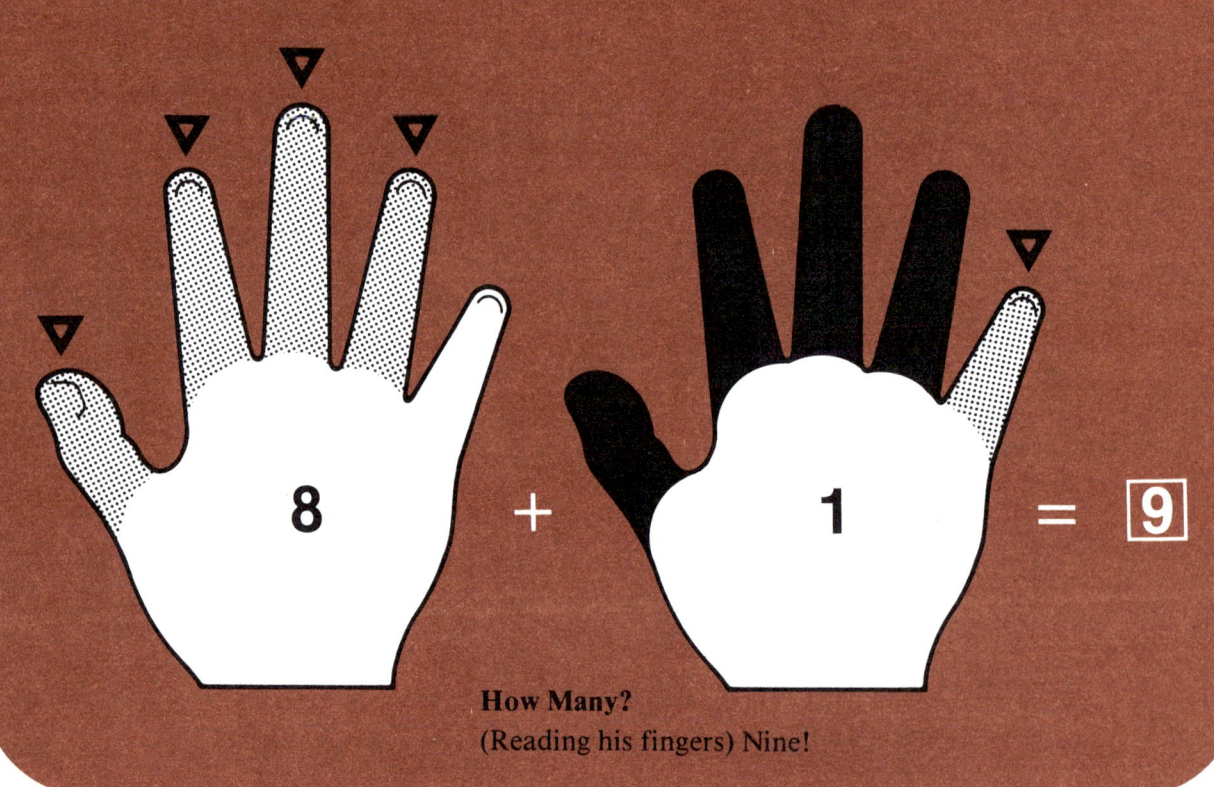

8 + 1 = 9

How Many?
(Reading his fingers) Nine!

Press
and
Clear

A Manipulative Shortcut

STOP

Do not proceed to this advanced method until you are certain that the following skills have been well established and can be performed without hesitation:

A. Recognition of all numbers from 1 to 99 and the ability to quickly press any number, on command, without hesitation.

B. Ability to add numbers using FIRST STAGE, UNIT-BY-UNIT progression ONLY. This skill must be mastered to include the Addition of numbers called orally. Repeated practice every day of these verbal exercises, adding "strings" of numbers up to sums of 99, must be continued —even when more advanced skills are being learned. This is comparable to the daily practice of musical scales when learning to play an instrument.

C. Ability to stop at any point in a calculation, quickly recognize and state the accumulated total and then continue with the Addition of more numbers (written or oral).

It is likely that the very young child will require several months of FIRST STAGE practice to be properly prepared for this advanced stage.

SLOW DOWN! Accuracy and the understanding of how numbers work are both more important than speed. Mastery of FIRST STAGE, UNIT-BY-UNIT CHISANBOP is the key to all future skills.

Press and Clear
A Manipulative Shortcut
(Right Hand)

All of the preceding calculations were done by Pressing fingers for one value and then Pressing more fingers to add a second value.

Now we introduce the "Press and Clear" concept which totally bypasses the need to add a number by Pressing one finger at a time in sequence.

Example: 1 + 4 = 5

Earlier, to have calculated this problem, we would:

Press One (First Finger) and then Press Four more by adding, one at a time, the Second, Third and Fourth Fingers. Then, finally, by Pressing the Fifth Finger (Thumb) and simultaneously Clearing the first four fingers. This was the simple and basic sequence from 1 to 5.

However, the "Press and Clear" concept obviates this single finger progression, as follows:

You've done so nicely learning to add numbers. Wouldn't it be fun if you were able to do this even faster? I'll show you what I mean. (Write 1 + 4 = □)

Here's an example we've done before.

(Use CHISANBOP fingering as you express it orally.) **One** (Press First Finger) **plus Four more.** (Starting your count again), **One** (Press Second Finger), **Two** (Press Third Finger), **Three** (Press Fourth Finger), **Four** (Press Fifth Finger and Clear other fingers) . . . **equals how many?**

Child: (Reading *your* fingers) Five!

Let's do it together and count out loud together. Ready? (Clear fingers) **Begin!**

Together:

One (Press First Finger) **plus Four** (Press fingers in sequence while counting). **One, Two, Three, Four.**

Parent: How Many?

Child: Five.

Now, keep your finger Pressed while I explain something to you. When we add the number Four to the number One, we get Five. But you can learn to add the Number Four to the number One much faster. We can skip all the fingers in the middle like this. Now watch closely. (Clear fingers.)

First I Press One—as I did before (Press First Finger to desk.)

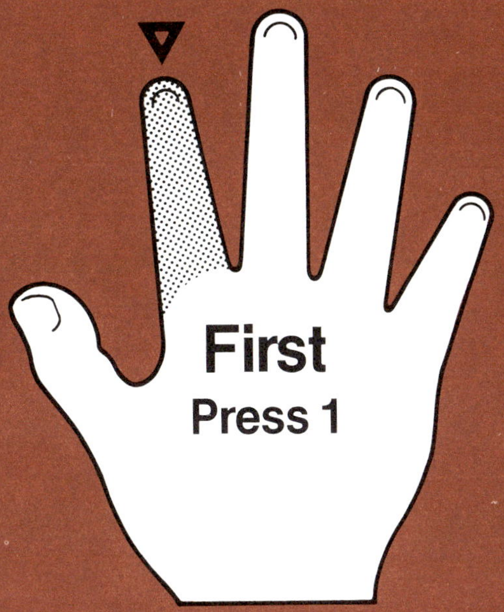

First
Press 1

Now . . . here's the trick for adding Four more. I'm going to Press Five and Clear One—at the same time.

See . . . we still end up with Five. Watch that again (Clear fingers). **First, I Press One.** (Press First Finger) **Then, I Press Five and Clear One at the same time.** (Do so) **Now. We're going to add One plus Four. Ready? Clear your fingers. Begin! Press One. Now we add Four more by Pressing Five and Clearing One.** (Repeat several times until established) **So you can see that from now on when we want to add Four and One, we can quickly Press Five and Clear One. Now this same trick works when we add Four and Two.**

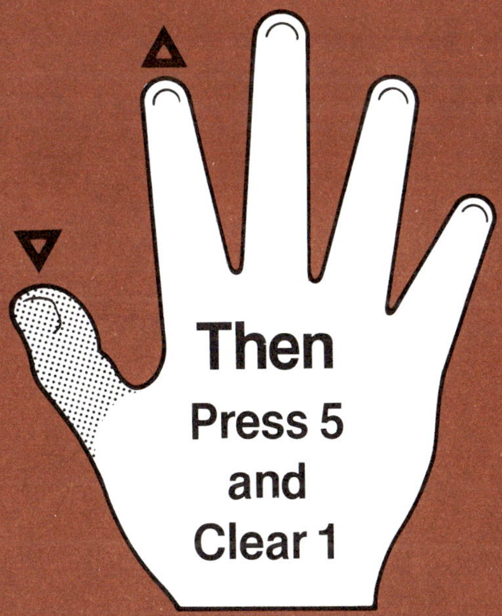

Continue this set with the following examples, each of which demonstrates that when we wish to add Four to the numbers 2, 3 or 4—there are not four fingers available, each with a value of One, which can be pressed in a single action. We conclude, however, that we reach the same end result by Pressing the available Fifth Finger (5) and Clearing the last single finger.

Then
Press 5
and
Clear 1

Simultaneously

2 + 4 = $\boxed{6}$

First, Press Two
(First and Second Fingers)
Next, Press Five
(Fifth Finger)
simultaneously Clearing One
(Second Finger).

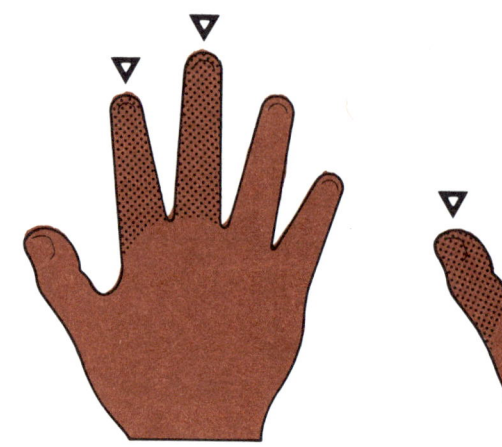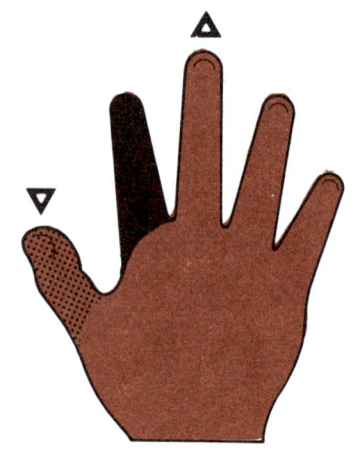

---

3 + 4 = $\boxed{7}$

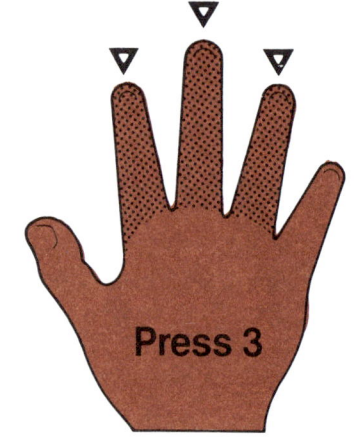

Press 3

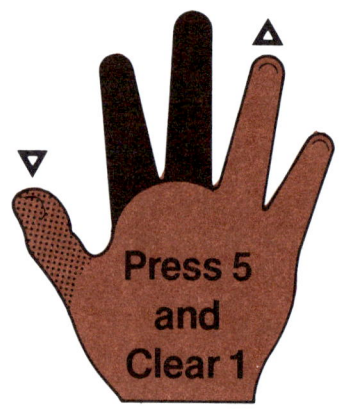

Press 5
and
Clear 1

Simultaneously

---

4 + 4 = $\boxed{8}$

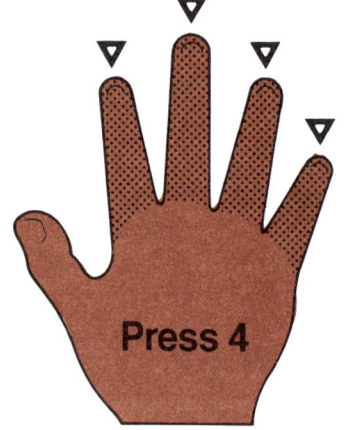

Press 4

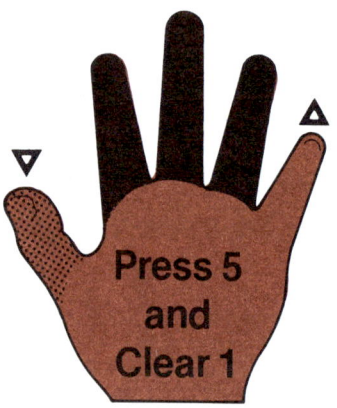

Press 5
and
Clear 1

Simultaneously

Just as we "Press 5 and Clear 1" as a shortcut to adding Four, the following set demonstrates similar Press and Clear substitutions, all of which are learned with dexterity through repetition.

2 + 3 = $\boxed{5}$

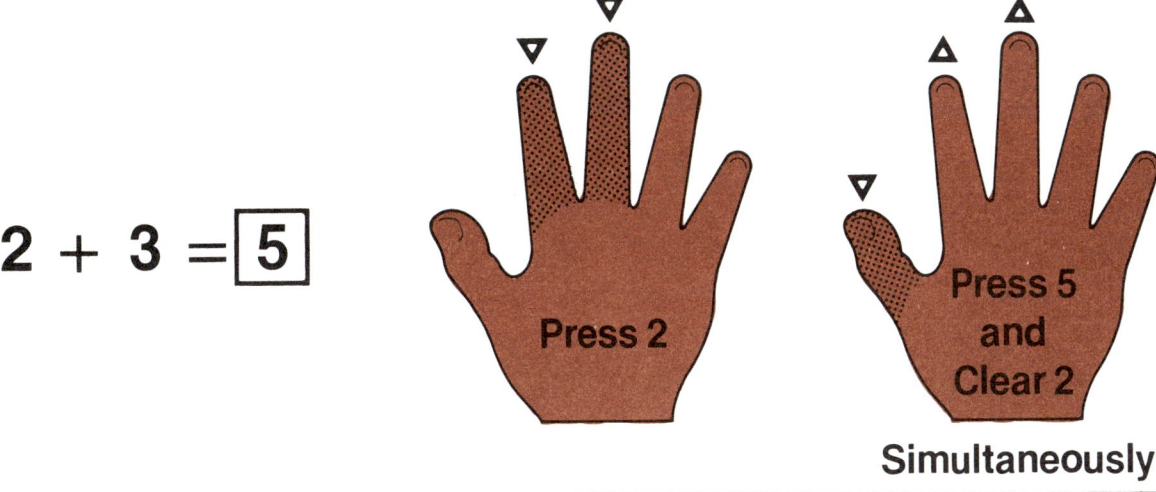

Press 2

Press 5 and Clear 2

Simultaneously

By first demonstrating through single finger progression that 2 plus 3 equals 5, one learns that the addition of 3 to any other number delivers the same result as the simultaneous addition of 5 and the Clearing of 2. Therefore, with CHISANBOP, the shortest route to a Plus 3 manipulation (assuming three adjacent fingers are unavailable) is by Pressing 5 and Clearing 2. Following are two more examples of a Plus 3 manipulation.

3 + 3 = $\boxed{6}$

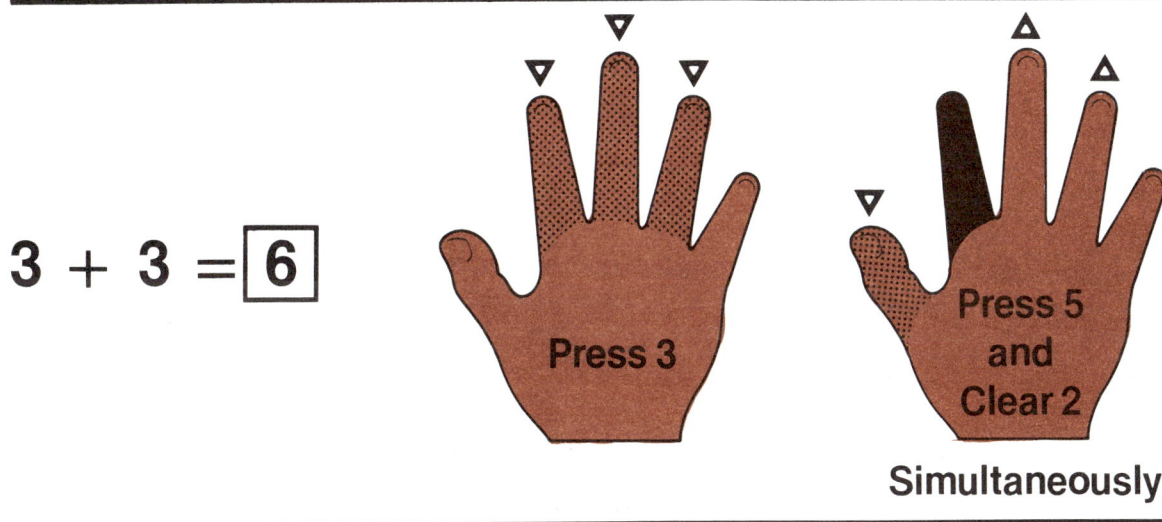

Press 3

Press 5 and Clear 2

Simultaneously

4 + 3 = $\boxed{7}$

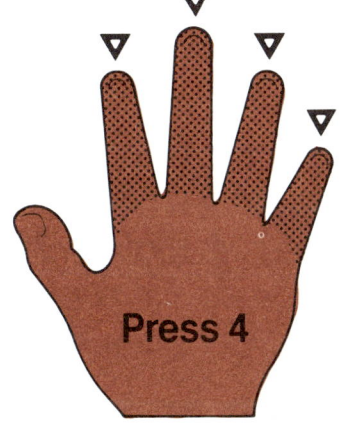

Press 4

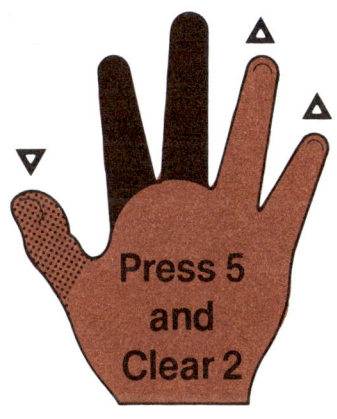

Press 5 and Clear 2

Simultaneously

Press and Clear
Both Hands

The next set of Press and Clear examples requires the manipulation of both hands.

$$5 + 5 = \boxed{10}$$

First, Press Five (Fifth Finger). Next, to add Five more, since it is impossible to Press an additional Five on the Right Hand, we perform a simultaneous Exchange, allowing us to use the Left Hand as follows:

Press Ten (Tenth Finger), which adds Five more than the problem requires. This excess is removed when we simultaneously Clear Five (Right Hand).

$$5 + 5 = \boxed{10}$$

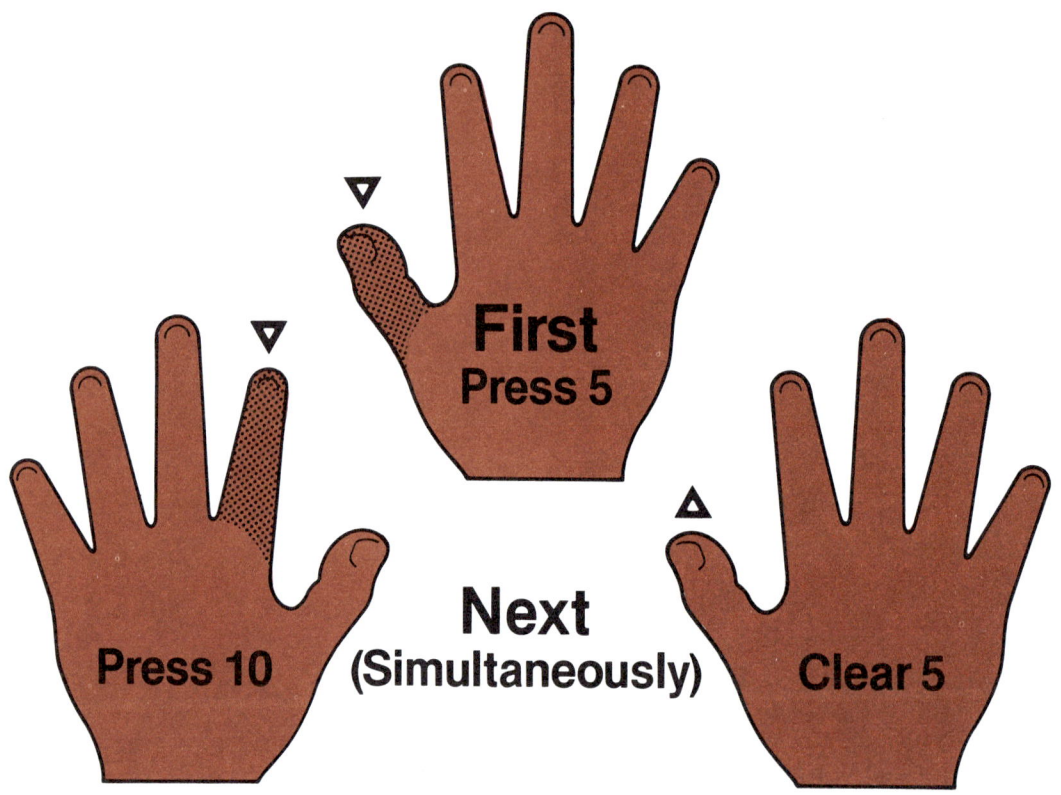

The following examples use the identical exchange in the process of adding 5. Each is to be practiced repeatedly until the procedure is established as an automatic response to the problem of adding 5 to the numbers 5, 6, 7, 8 and 9.

$$6 + 5 = \boxed{11} \qquad 8 + 5 = \boxed{13}$$
$$7 + 5 = \boxed{12} \qquad 9 + 5 = \boxed{14}$$

Adding 6

This set of examples presents a similar Exchange to the previous set. However, it requires an extra manipulation. Namely, when the Exchange occurs, Clearing the Fifth Finger for the Tenth Finger (which would only yield 5) we must now Press One more on the Right Hand, as follows:

$$\textbf{5} + \textbf{6} = \boxed{\textbf{11}}$$

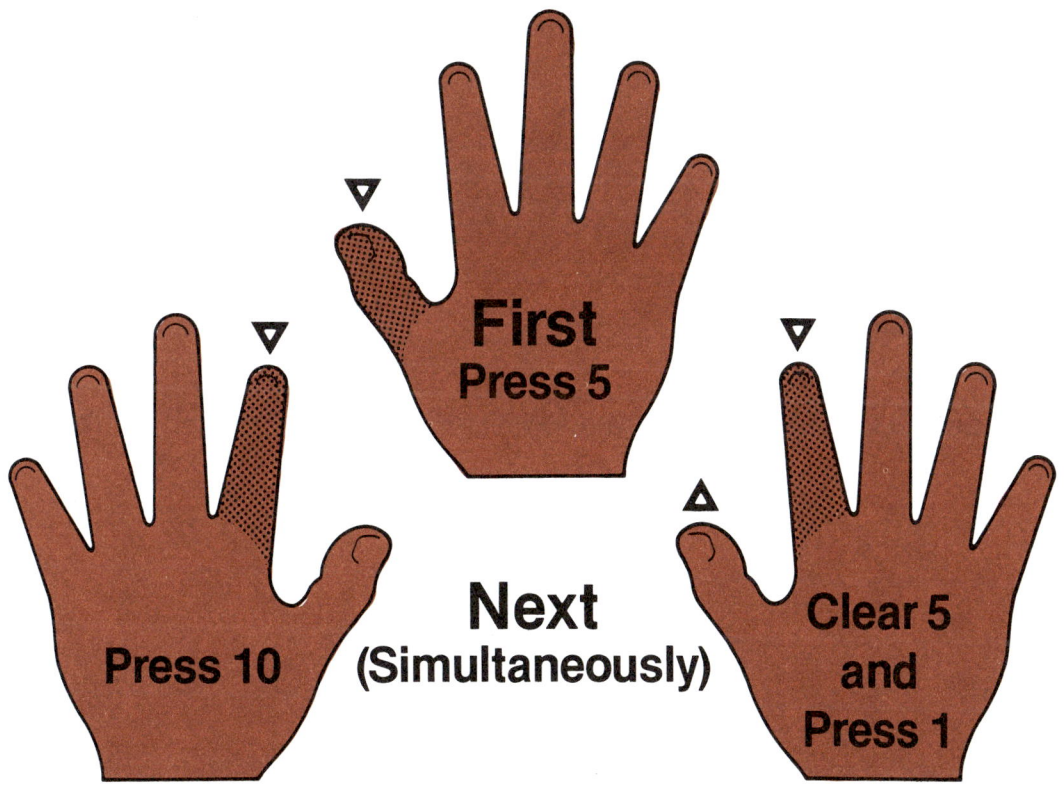

First Press 5

Press 10 **Next** (Simultaneously)

Clear 5 and Press 1

Continue with the following examples which use the identical exchange in the process of adding 6. Each is to be practiced repeatedly until the procedure is established as an automatic response to the problem of adding 6 to the numbers 5, 6, 7 and 8.

$$6 + 6 = \boxed{12} \qquad 7 + 6 = \boxed{13} \qquad 8 + 6 = \boxed{14}$$

Adding 7

The manipulation for these examples is the same as for adding 6, except that when the Exchange occurs, Clearing the Fifth Finger for the Tenth Finger and Pressing One more (to yield 6) we must now Press Two more on the Right Hand, as follows:

$$5 + 7 = \boxed{12}$$

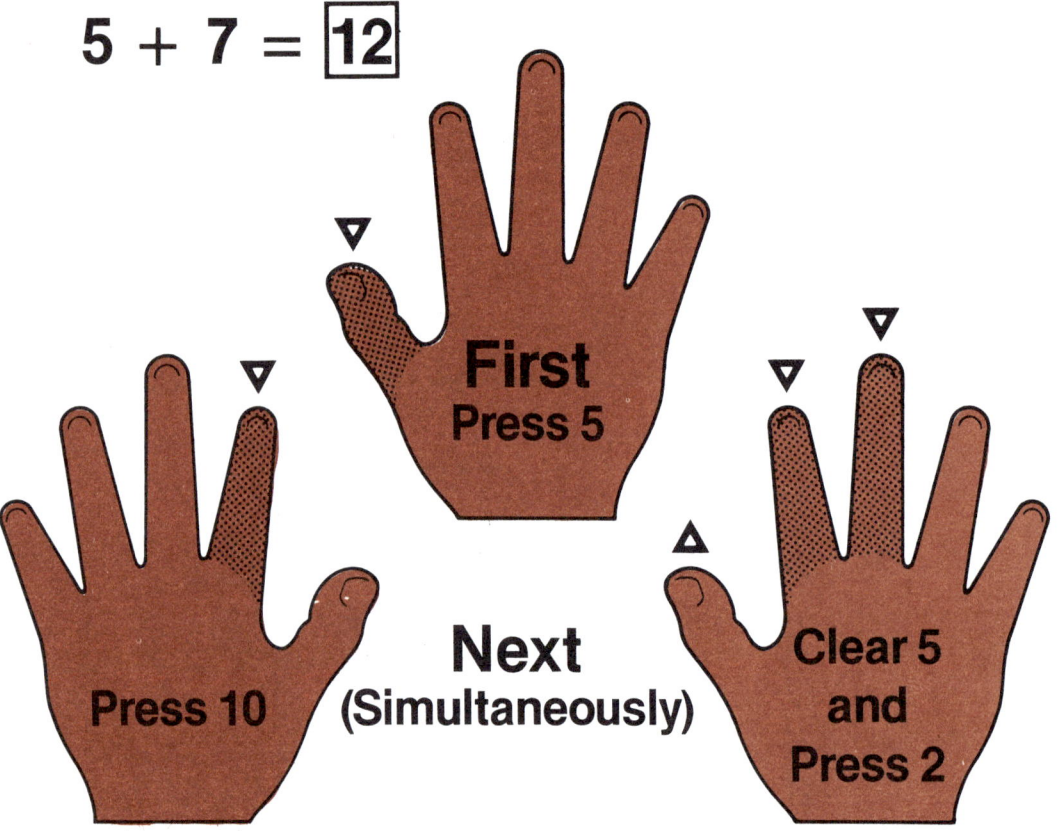

First
Press 5

Next
(Simultaneously)

Press 10

Clear 5
and
Press 2

Continue with the following examples which use the identical Exchange in the process of adding 7. Each is to be practiced repeatedly until the procedure is established as an automatic response to the problem of adding 7 to the numbers 5, 6 and 7.

$$6 + 7 = \boxed{13} \qquad 7 + 7 = \boxed{14}$$

Adding 8 or 9

These examples follow the identical pattern first established for adding 6 or 7, except that to add 8 to the numbers 5 or 6, we Clear 5 and Press 3 on the Right Hand while Pressing 10 on the Left. To add 9 to the number 5, we Clear 5 and Press 4 on the Right Hand while Pressing 10 on the Left. Following are the only examples requiring these Exchanges.

$$5 + 8 = \boxed{13} \qquad 6 + 8 = \boxed{14} \qquad 5 + 9 = \boxed{14}$$

Adding 9 To Other Numbers

This set of examples presents a comparatively simple Press and Clear manipulation. With the knowledge that 9 is the same as 10 less 1, we accomplish the plus 9 manipulation as follows:

$$1 + 9 = \boxed{10}$$

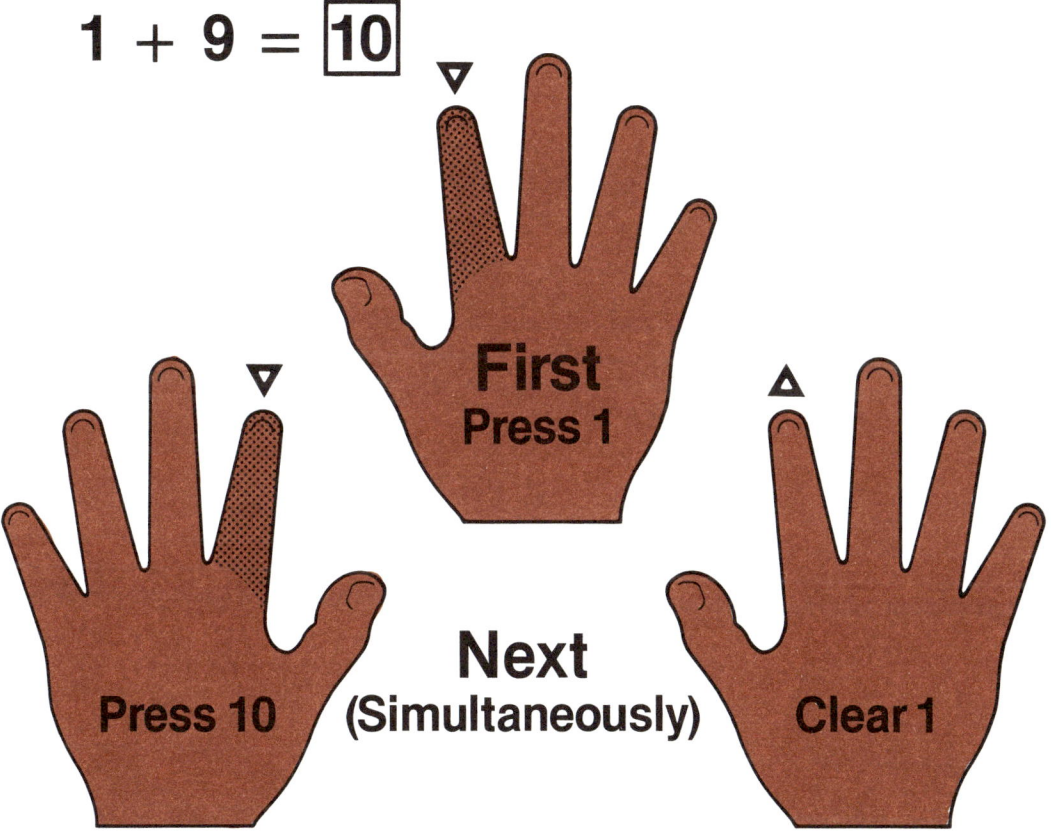

First Press 1

Press 10

Next (Simultaneously)

Clear 1

Continue with the following examples which use the identical Exchange in the process of adding 9. Each is to be practiced repeatedly until the procedure is established as an automatic response to the problem of adding 9 to the numbers 1, 2, 3, 4, 6, 7, 8 and 9.

$$2 + 9 = \boxed{11} \qquad 6 + 9 = \boxed{15} \qquad 8 + 9 = \boxed{17}$$
$$3 + 9 = \boxed{12} \qquad 7 + 9 = \boxed{16} \qquad 9 + 9 = \boxed{18}$$
$$4 + 9 = \boxed{13}$$

Adding 8 To Other Numbers

This set of examples follows the same pattern as the previous set for adding 9. With the knowledge that 8 is the same as 10 less 2, we accomplish the plus 8 manipulation as follows:

$$2 + 8 = \boxed{10}$$

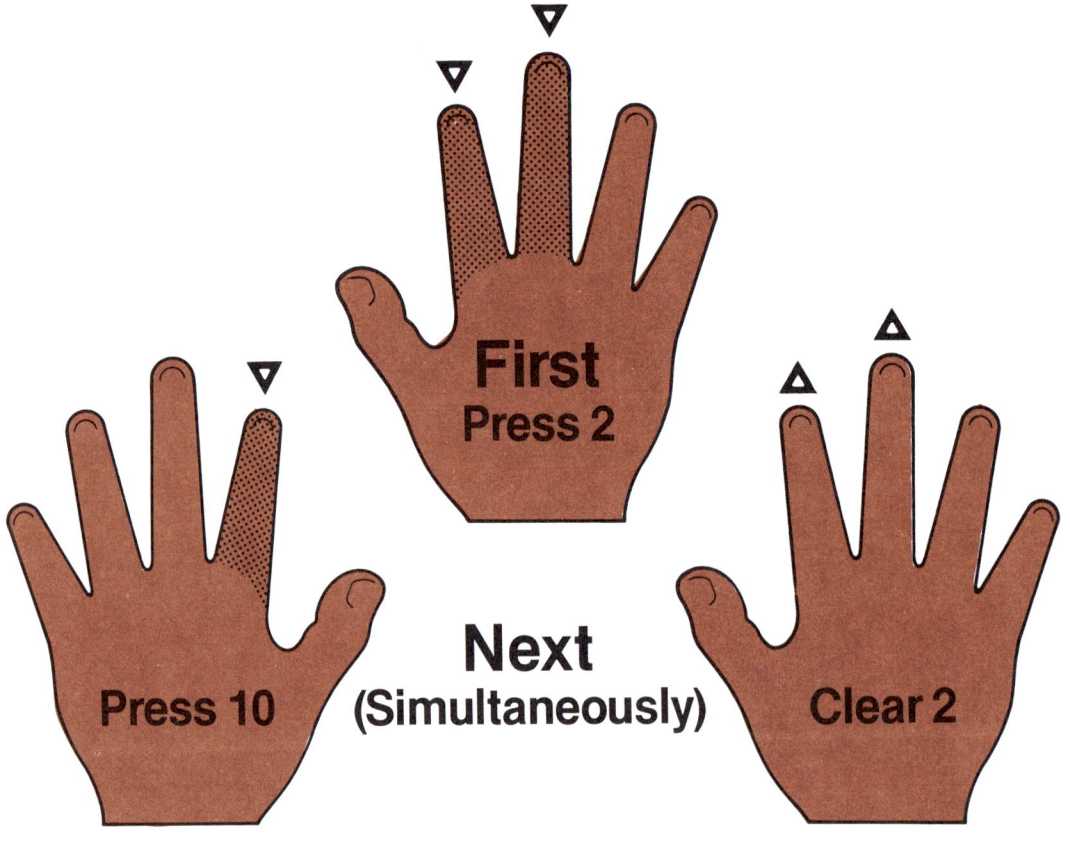

First
Press 2

Next
(Simultaneously)

Press 10

Clear 2

Continue with the following examples which use the identical Exchange in the process of adding 8. Each is to be practiced repeatedly until the procedure is established as an automatic response to the problem of adding 8 to the numbers 2, 3, 4, 7, 8 and 9.

$3 + 8 = \boxed{11}$ $8 + 8 = \boxed{16}$
$4 + 8 = \boxed{12}$ $9 + 8 = \boxed{17}$
$7 + 8 = \boxed{15}$

Adding 7 To Other Numbers

This set of examples follows the same pattern as the previous set for adding 8. With the knowledge that 7 is the same as 10 less 3, we accomplish the plus 7 manipulation as follows:

$$3 + 7 = \boxed{10}$$

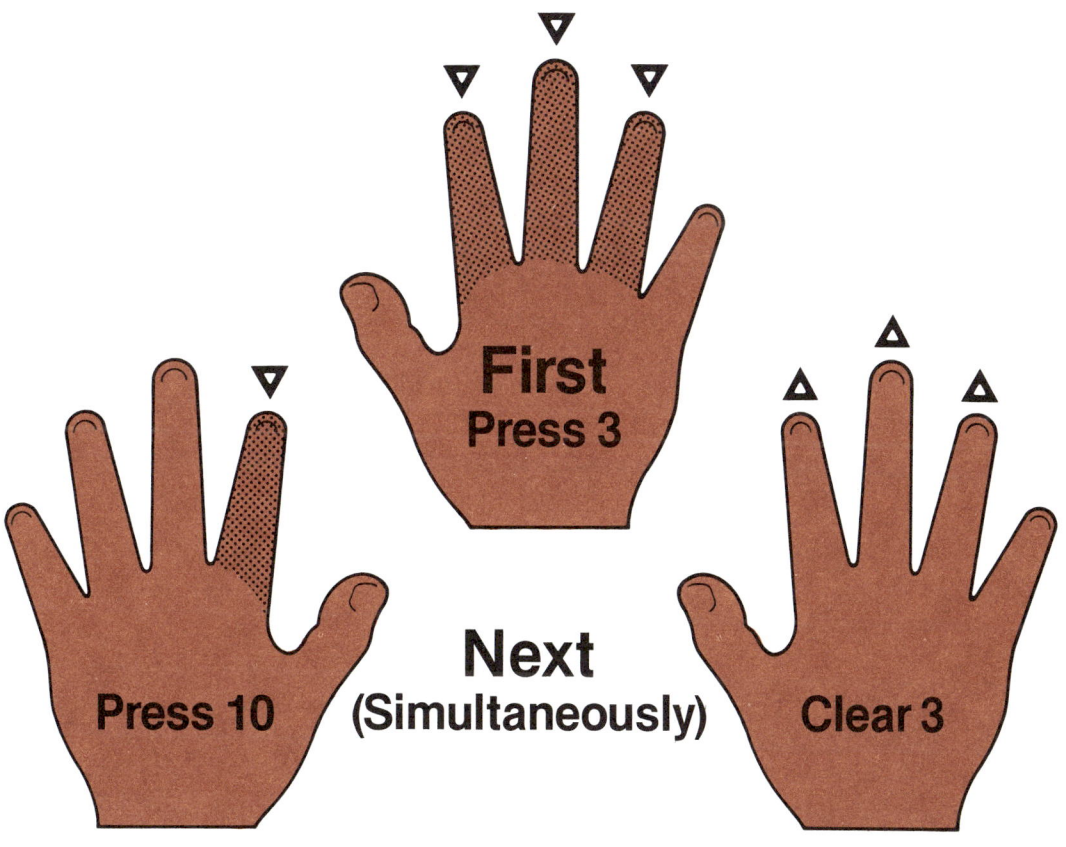

First
Press 3

Press 10

Next
(Simultaneously)

Clear 3

Continue with the following examples which use the identical Exchange in the process of adding 7. Each is to be practiced repeatedly until the procedure is established as an automatic response to the problem of adding 7 to the numbers 3, 4, 8 and 9.

$$3 + 7 = \boxed{10} \qquad 8 + 7 = \boxed{15}$$
$$4 + 7 = \boxed{11} \qquad 9 + 7 = \boxed{16}$$

Adding 6 to Other Numbers

This set of examples follows the same pattern as the previous set for adding 7. With the knowledge that 6 is the same as 10 less 4, we accomplish the plus 6 manipulation as follows:

$$4 + 6 = \boxed{10}$$

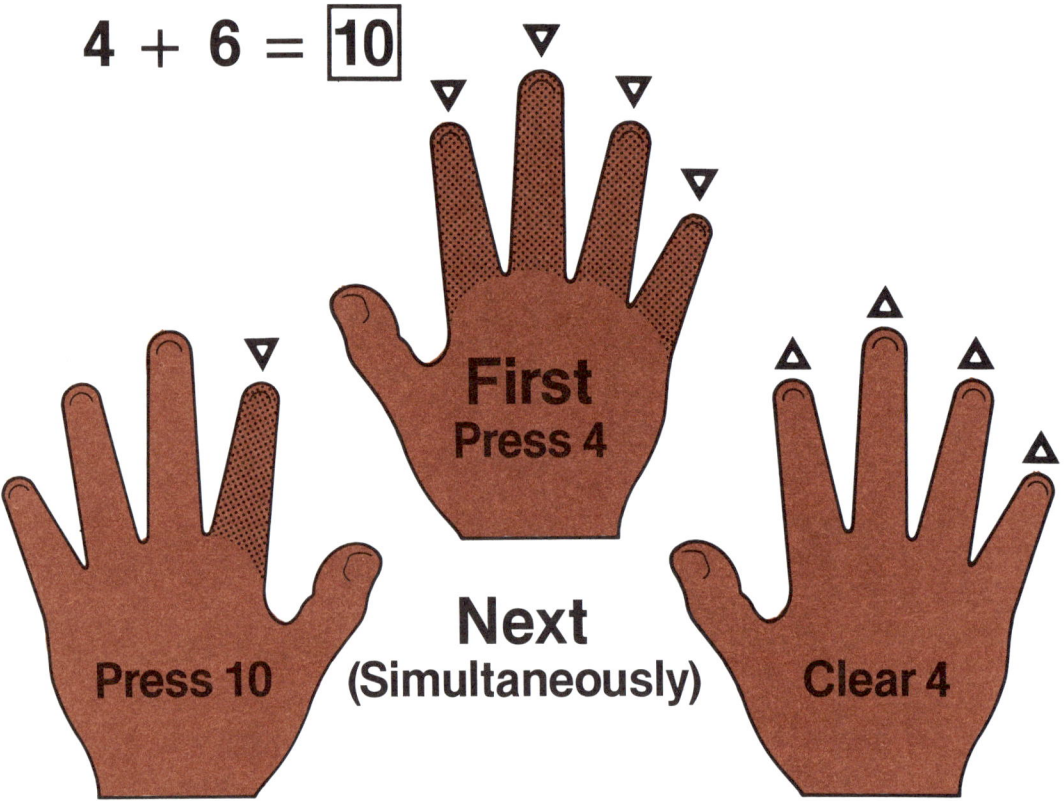

First
Press 4

Next
(Simultaneously)

Press 10

Clear 4

There is only one other example using the identical Exchange in the process of adding 6.

$$9 + 6 = \boxed{15}$$

Practice both examples until the procedure is established as an automatic response to the problem of adding 6 to the numbers 4 and 9.

Adding 4 To Other Numbers

This set of examples follows the reciprocal pattern as the previous set for adding 6. As 6 was the same as 10 less 4, so is 4 the same as 10 less 6. We accomplish the plus 4 manipulation as follows:

$$6 + 4 = \boxed{10}$$

First
Press 6

Next
(Simultaneously)

Press 10

Clear 6

Continue the following examples which use the identical Exchange in the process of adding 4. Each is to be practiced repeatedly until the procedure is established as an automatic response to the problem of adding 4 to the numbers 6, 7, 8 and 9.

$$7 + 4 = \boxed{11} \qquad 8 + 4 = \boxed{12} \qquad 9 + 4 = \boxed{13}$$

Adding 3 to Other Numbers

This set of examples follows the same pattern as previous sets, inasmuch as 3 is the same as 10 less 7, so we accomplish the plus 3 manipulation as follows:

$$7 + 3 = \boxed{10}$$

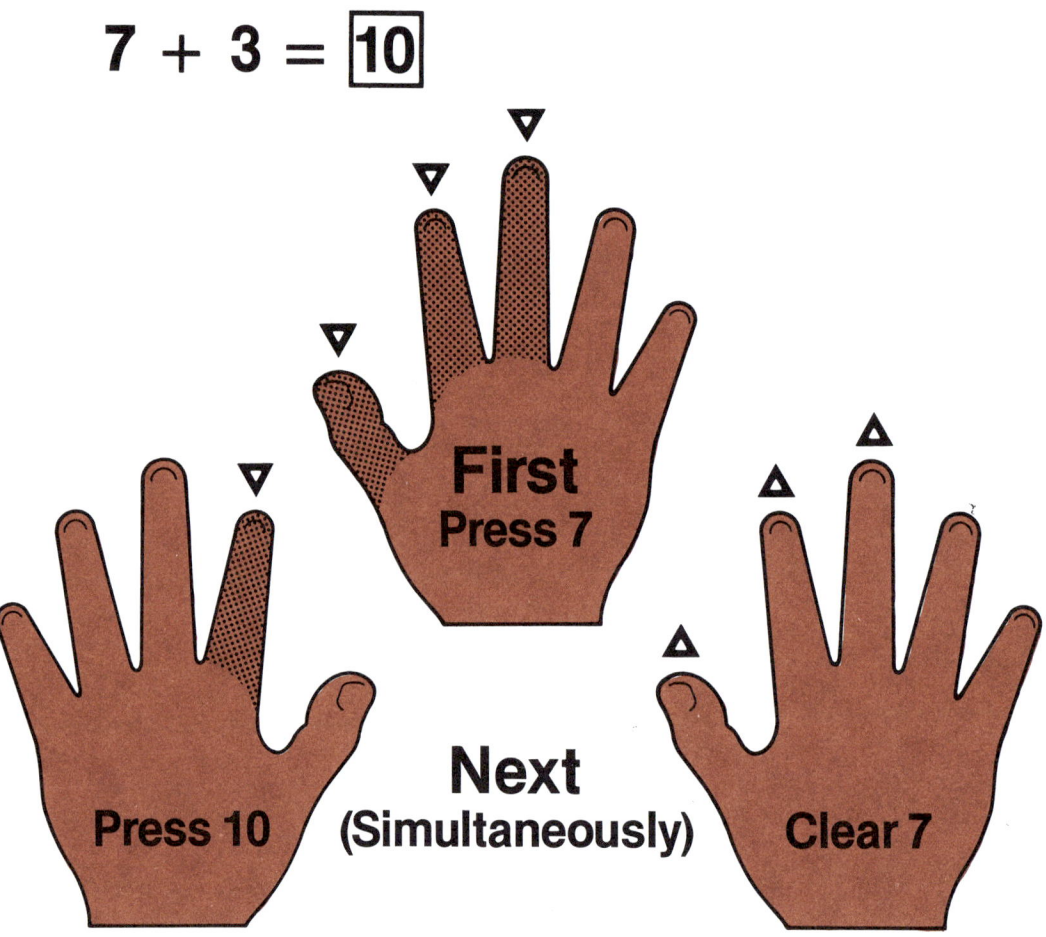

First
Press 7

Next
(Simultaneously)

Press 10

Clear 7

Continue with the following examples which use the identical Exchange in the process of adding 3. Each is to be practiced repeatedly until the procedure is established as an automatic response to the problem of adding 3 to the numbers 7, 8 and 9.

$$8 + 3 = \boxed{11} \qquad 9 + 3 = \boxed{12}$$

Adding 2 to Other Numbers

This is the final set which follows the pattern of the previous sets, inasmuch as 2 is the same as 10 less 8, so we accomplish the plus 2 manipulation as follows:

$$8 + 2 = \boxed{10}$$

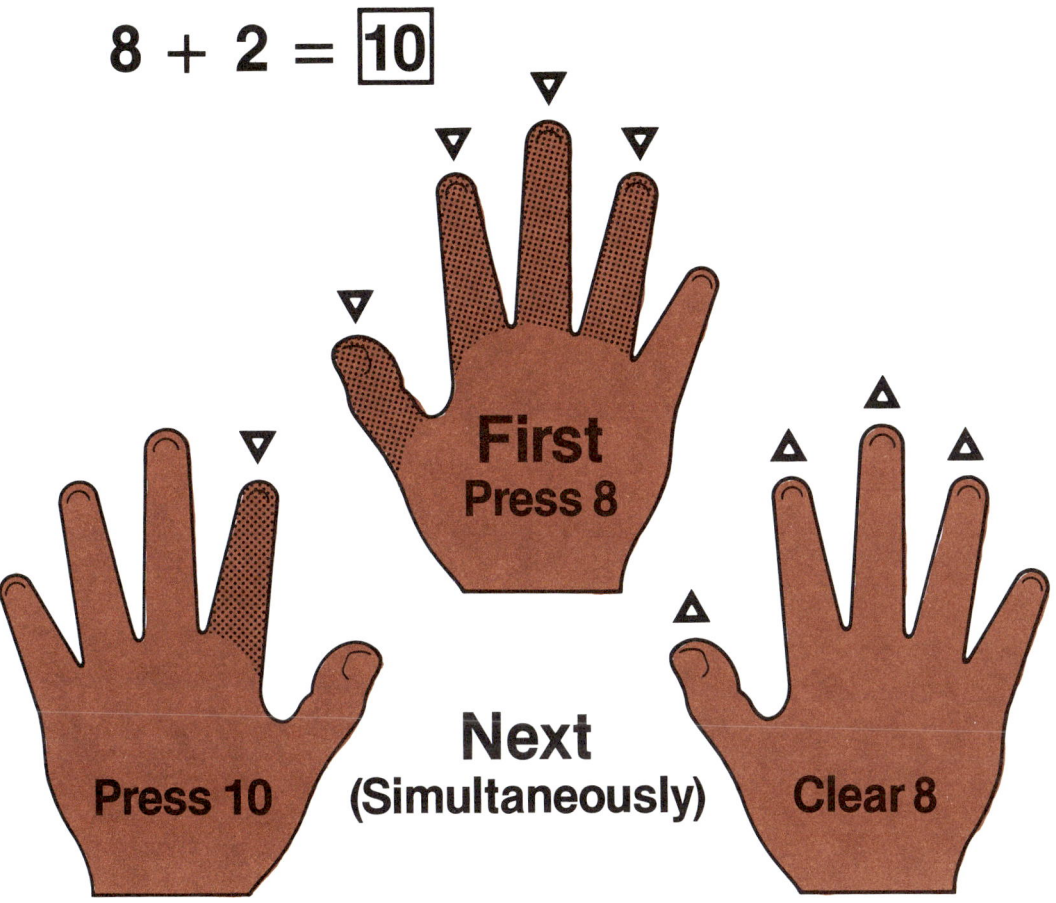

There is only one other example using the identical Exchange in the process of adding 2.

$$9 + 2 = \boxed{11}$$

Practice both examples until the procedure is established as an automatic response to the problem of adding 2 to the numbers 8 and 9.

Having acquired the skills for
performing the preceding manipulations,
the following calculations can
now be addressed.

Multi-Digit Addition

Algebra Addition

Multi-Digit Addition

An outstanding asset of the CHISANBOP method appears clearly when one wishes to add a column of numbers, for at any point in the calculation the total is maintained and can be read on one's fingers. This is an advantage over conventional procedures which require mental retention of the accumulated total.

EXAMPLE:	MANIPULATION	LEFT	RIGHT	ACCUMULATED TOTAL
3	Press 3			3
4	Press 5, Clear 1			7
2	Press 2			9
1	Press 10, Clear 9			10
3	Press 3			13
+ 2	Press 5, Clear 3			15

In a more advanced stage, we would combine sets of two numbers and Press their combined value. In the above example, only three manipulations would occur:

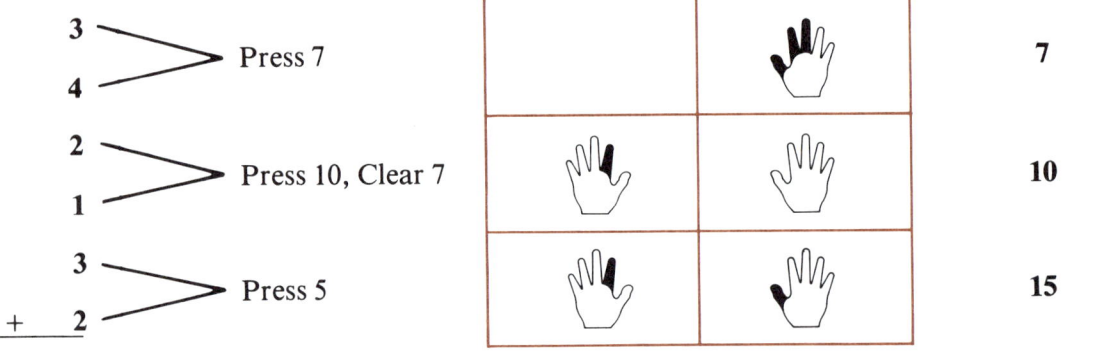

3 4	Press 7		7
2 1	Press 10, Clear 7		10
3 + 2	Press 5		15

Adding More than One Column

EXAMPLE:

```
        2   2 2
        2 9 6
        8 3 4
        5 6 1
    +   7 5 8
    ─────────
        2,449
```

The method employed here is to add the right column with the standard manipulation. Finding our total to be 19, our fingers are positioned:

Write the Number 9 below the line in conventional fashion, while *keeping the Tenth Finger Pressed*. Now, observing the single finger Pressed, this is transferred to a single finger (First) on the Right Hand. This serves to "hold" the carry-over number. Then, with the First Finger already Pressed, begin the addition of the second column. On reaching the total 24, we enter 4 below the line, observe the two Pressed fingers on the Left Hand and transfer these to the Right Hand (First and Second Fingers). Now we are prepared to add the third column, having carried over the number 2. On reaching the total 24, we enter that entire number.

Therefore, with any column totalling up to 99, we write the Units number— which is always on the Right Hand—below the line, and transfer the number of Left Hand Fingers to the Right Hand as a carry-over.

In the early stages of Multi-Digit addition, the child should enter the carry-over number above the next column to be added. The Home Study First Stage Addition Workbook provides boxes in which carry-over numbers can be entered.

LEFT-HANDED CHILDREN REQUIRE A MODIFICATION

In the above example, the LEFT-HANDED CHILD, having registered the Units column total on his fingers thusly: 19

will first write the LEFT-HAND Carryover number (1) above the Tens column. Then he proceeds to enter the Right-Hand Number (9) below the Units column. Addition of the Tens column follows, yielding: 24

The procedure is repeated, first entering the LEFT-HAND Carryover number (2) and then the RIGHT-HAND Number (4).

Ultimately, written entry of the Carryover is abandoned. Instead, child will enter the RIGHT-HAND Number below the line, while transferring the LEFT HAND NUMBER OF FINGERS to the RIGHT HAND. Having done so, addition of the next column can then begin.

Algebra Addition

Having acquired the skills to perform Addition with CHISANBOP, the examples of simple Algebra Addition are relatively easy to comprehend:

(Write on paper)

$$1 + 1 = 2$$

Remember when we learned to do this simple example? (Press Fingers while orally stating example)
One (Press First Finger) **plus One More**
(Press Second Finger) **equals Two.**
(Draw a vertical line through the "equals" sign:)

$$1 + 1 \neq 2$$

This means that both sides of this example have to balance . . . like a see-saw. If you put too much on this side (point to $1 + 1$) **or on this side** (point to 2), **it just won't balance. And this "equals" sign** (write it again) **means that everything on this side is the same as everything on the other side. First you Press One.** (Observe) **Now, Press One more.** (Observe) **How Many?**

Child: Two!

Right! So this side that says "1 + 1" is the same as this side that says "2". They balance.
Now, let's play a little trick on ourselves. Watch. Write:

$$1 + \square = 2$$

Before (point to $1 + 1 = 2$ example) **we said "One plus One equals How Many?" Now we say** (pointing) **One plus How Many equals Two?**
Remember, now. Everything on *this* side of the equals sign (point to $1 + \square$) **must balance with everything on this side** (point to 2).
First, I Press One. (First Finger) **Now I have to find out How Many more will add up to Two** (Point to $\square$) **Watch my fingers and tell me when I reach Two. I Press One More.** (Second Finger)

Child: Now!

Right! Now I have Two. So, How Many did I add to this One to add up to Two?

Child: One!

Good. Now let's try one that's a little harder.

Write:

$$1 + \square = 3$$

Remember now, everything on this side of the equals sign (point to 1 + □) **must balance with everything on this side** (point to 3).

First, I Press One (First Finger). **Now I have to find out How Many more** (point to □) **will add up to 3** (point to 3). **Watch my fingers and tell me when I reach Three. One** (Press Second Finger), **Two** (Press Third Finger).

Child: Now!

Good. (Repeat manipulation.) **First I Pressed One and then I Pressed Two more to make my fingers equal Three. So. How Many go here?** (Point to □)

Child: Two!

Good. (Write 2 in □) **Now we'll try it together.**

Continue the examples in Workbook, always emphasizing the need to balance on both sides of the equals sign. Be cautious in the beginning stage that the How Many count always begins with "One". Also stress the need to "read your fingers" as the count progresses to determine when to stop, for the oral count will always differ from the accumulated finger count. Progress, as with Basic Addition, to more advanced stages, abandoning finger-by-finger progression. For example:

$$1 + \square = 6$$

Press One (First Finger).

Press Five More (Fifth Finger) observing total finger accumulation of Six. Enter 5 in the □.

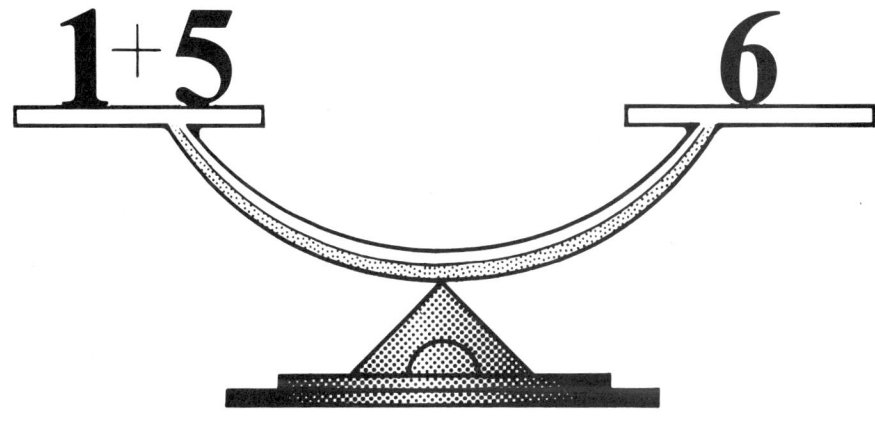

Correct CHISANBOP position of both hands when all fingers are pressed. Note that there is a relaxed, natural arch and that only the finger tips make contact. There is a strong similarity bewtween the CHISANBOP position and that which is assumed for playing piano or fingering a typewriter.

Conclusion

Home Study CHISANBOP—Addition, used in conjunction with the corresponding First Stage Addition Workbook, provides the foundation for all subsequent CHISANBOP study. It is recommended that the First Stage Addition Workbook be purchased inasmuch as it is organized according to the sequence of techniques outlined in this book. The following sample pages provide an idea of the format used in the 80-page Workbook. Workbook examples are identical to those used by schools which have adopted CHISANBOP as part of their curriculum.

CHISANBOP ENTERPRISES can provide qualified Teacher-Instructors to schools for preparation of classroom activities. Workshops necessary for Teacher Certification require approximately 30 hours of training to assure comprehensive knowledge of Chisanbop principles and techniques.

Schools desiring workshops for teachers should direct all inquiries to CHISANBOP ENTERPRISES, INC., P.O. Box 99, Mount Vernon, New York, 10551.

Totals up to 4

SAMPLE WORKBOOK PAGE

0	0	0	0
+ 1	+ 2	+ 3	+ 4

1	1	1	2
+ 1	+ 2	+ 3	+ 2

0 + 1 = ☐

0 + 2 = ☐

0 + 3 = ☐

0 + 4 = ☐

1 + 1 = ☐

1 + 2 = ☐

1 + 3 = ☐

2 + 2 = ☐

ACTUAL WORKBOOK SIZE 8½" x 11"

2	0
0	4
+ 2	+ 0

3	1
0	1
+ 1	+ 2

1	0
2	0
+ 1	+ 1

1	2
0	1
+ 1	+ 0

$$1 + 2 = \boxed{}$$

$$0 + 1 = \boxed{}$$

$$1 + 1 = \boxed{}$$

$$0 + 2 = \boxed{}$$

$$1 + 3 = \boxed{}$$

$$0 + 3 = \boxed{}$$

$$2 + 2 = \boxed{}$$

$$0 + 4 = \boxed{}$$

● ● ● + ● = $\boxed{}$

◆◆ + ◆◆ = $\boxed{}$

✳ + ✳ = $\boxed{}$

SPEED

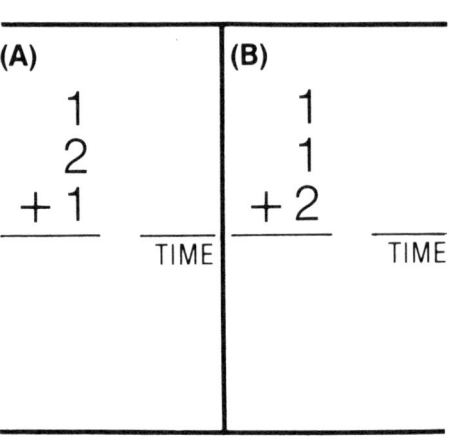

(A)
$$\begin{array}{r} 1 \\ 2 \\ + 1 \\ \hline \end{array}$$

(B)
$$\begin{array}{r} 1 \\ 1 \\ + 2 \\ \hline \end{array}$$ TIME ___

TIME ___

(C)
$$\begin{array}{r} 2 \\ 0 \\ + 1 \\ \hline \end{array}$$
$$\begin{array}{r} 1 \\ 1 \\ + 0 \\ \hline \end{array}$$ TIME ___

(D)
$$\begin{array}{r} 0 \\ 1 \\ + 0 \\ \hline \end{array}$$
$$\begin{array}{r} 2 \\ 1 \\ + 1 \\ \hline \end{array}$$

$$\begin{array}{r} 1 \\ 0 \\ + 1 \\ \hline \end{array}$$
$$\begin{array}{r} 1 \\ 1 \\ + 1 \\ \hline \end{array}$$ TIME ___